Focus First on Service:

The Voice and Face of Your Utility

Focus First on Service:

The Voice and Face of Your Utility

Second Edition

Maria DeChellis
Jim Ginley
Tarja Nummela

American Water Works Association

Disclaimer
The authors, contributors, and publisher do not assume responsibility for the validity of the content or any consequences of their use. In no event will AWWA be liable for direct, indirect, special, incidental, or consequential damages arising out of the use of information presented in this book. In particular, AWWA will not be responsible for any costs, including, but not limited to, those incurred as a result of lost revenue. In no event shall AWWA's liability exceed the amount paid for the purchase of this book.

ISBN: 978-1-64717-291-6 ISBN, electronic: 978-1-61300-811-9
https://doi.org/10.12999/AWWA.20629ed2

Sr. Manager — Product Acquisition & Development: Geoffrey S. Shideler
Manager — Publishing Operations: Gillian Wink
Specialist — Copyright and Permissions: Peggy Tyler
Director of Publishing: John Fedor
Technical Editor: Maureen Peck
Technical Editing Review: Anne Lanctôt
Cover Design: Michael Labruyere
Cover Images: Pixel-Shot, HY-DP, Dewitt, Opat Suvi, Leny Studio, Monkey Business Images, Kmpzzz and pranshiko sanovic/Shutterstock.com
Production: Innodata

American Water Works Association

Contents

List of Figures

List of Tables

Gratitude

The authors would like to thank all of the clean water professionals who are the backbone of providing water services all over the world. We would especially like to thank the water professionals that we have worked with over these last years, who, while always our colleagues and mentors, have become friends, namely,

Steve Miller, Greenville Water
Kelly Dearing-Smith, Louisville Water
Kathy Koch, formerly of the Portland Water Bureau
Shannon Frye, Metro Water Services, Nashville
Mike Aragon, Denver Water
Chima Chijoke, Southern Company
Heather Pressley, Charlotte Water
Sheila Pressley, formerly of Jacksonville Electric Authority
Nicole Bates, Mt. Pleasant Water Works
Madeleine Driscoll, Hazen and Sawyer
Dave Vogel, Louisville Water
Yvonne Moore-Jackson, Baltimore Department of Public Works
Vanessa Garrett-Ingram, formerly of Baltimore Department of Human Resources
Hanan Dilou, National Water Company, Kingdom of Saudi Arabia
Hamza Triki, National Water Company, Kingdom of Saudi Arabia
Christina Martinez, Oracle
Simon Warren, City of Durham
Travis Thompson, Denver Water
Jimmy Luthye, Denver Water
Alex Hayner, Charlotte Water
Kayla Sanon, Richmond (VA) Department of Public Utilities
Cheryl Muldrow, Town of Wingate
Zachary Tarlton, Union County Water
Jennifer Corbett, Union County Water
Amanda Jones, Womxn in Water
Amy McCaskill, Union County Water
Kate Rowland, Oracle
Comune di Pacentro (AQ), Italy

A special thank you to AWWA staff, namely:

Geoffrey Shideler
Maureen Peck
Anne Lanctôt

For the professionals who have given their time and effort in providing feedback, reviewing, and being open and challenging to produce this labor of love.

Heidi Hackett, Cape Fear Public Utility Authority
Douwe Busschops, H2O America
Dennis DeMera, Las Vegas Valley Water

To the employers who gave us the freedom to take on this work:

The City of Tempe, a small city with big ideas and even bigger drive for excellence

Oracle
SL-Serco

And most importantly to our families: David Ehrenpreis, Nayyar Bhatti, Carlo and Anna Maria De Chellis, and our children who have inspired us to see the impact and importance of clean water for all: Alessandra and Jennie Belle Sisk and Amalia Bhatti.

Chapter 1

Customer Service: Role Versus Responsibility

Even if customers can't choose their water or wastewater provider, they do compare the service provided by your utility with the best experience they've had anywhere else.

> "A ship in harbor is safe, but that is not what ships are built for."—John Augustus Thredd

Your utility was not designed to be a mystery to people; it was built to provide a life-sustaining service to a community. People don't compare you with their optional, nonessential services; they view their need for you as they do food, medical services, and the air they breathe. This makes the responsibility of customer service at your utility even more essential and possibly challenging, as you must address your entire service population likely without the plentiful budgets of private organizations. It also means that government-service trust levels and economic factors all sway the perception of the service you provide, right or wrong.

"So, to drink tap water is to trust government (Walton 2022)," says Manny Teodoro, professor at the University of Wisconsin–Madison, a distinguished researcher and advocate for water affordability.

That invisible comparison sets the bar for our teams and our culture.

Regardless of your role in a utility, you know that customer service is not a desk, a counter, or a call queue; it's a responsibility that lives across the organization. From field service and service billing to construction and communication, service is a shared duty and a career calling, centered on helping people understand, navigate, and trust the essential services they rely on every day.

Three truths guide much of what you will read in this book.

- **Service is a decision that has impacts, both positive and negative.** Clear choices about policy, training, tools, and staffing create predictable, helpful experiences.

- **Consistency across touchpoints builds reputation.** All channels of communication should be considered when thinking through any process. Phone, field engagement, written correspondence, digital, mobile applications, email, TV, and social media should feel like one utility voice.
- **Culture determines how your utility shows up for customers each day.** It's not up to just one department to provide good customer service.

If your job involves dealing with customers, you may feel like you're a therapist, advisor, accountant, regulator, attorney, and neighbor, often on the same day. Sure, we would all love water and wastewater services to be free for everyone, but producing safe, reliable drinking water and sanitary sewer service requires money and expertise. Bridging that gap with clarity and care is the heart of this work.

What you can expect over the course of this book is information about weaving together a culture of customer service that considers customer needs, humane communication, and ever-changing technology.

FROM DECISION TO DISCIPLINE

Turn the three truths into daily practice with these moves.

Design Policies for People

Translate rules into plain language. Many water regulations were written decades ago, and it has been a long time since the language in them has been updated. This makes much of the text largely unreadable for the general population because it uses a mix of jargon and industry-specific terminology.

For example, "Install an approved backflow prevention assembly within the time frame required by the city and in all instances within thirty (30) days."

> *What is the time frame? This is a city document; is it 30 days or is it another requirement that is found in a different document?*

Another example: "Should it be desired to discontinue the use of all water for a period of not less than fifteen days turned off and turned on again without charge; provided, however, no remission of rates will be made for a period of less than fifteen days."

> *With multiple uses of "no" and "not," the time frame seems unclear, and not all people are familiar with terms like the "remission of rates."*

Action: Review and read through your regulations and all correspondence, not just by management, but by community members of your service territory. Publish timelines that are visible and explain what customers can expect for common processes in simple terms (e.g., service start or stop, leaks, payment plans, claims). Every communication should be clear about what expectations are involved. Test your letters, emails, and texts with some actual customers before fully launching.

The book *Wankernomics* (2025) from the comedy duo James Schloeffel and James Firth pokes fun at the jargon that is used throughout businesses. Expressions such as "But does it ladder up to the overarching purpose statement?" can be word salad within the corporate world. Before you laugh, think about whether your regulations or process documents have a similar sentiment, and make the change if they do.

EQUIP EVERY ROLE

There is no end to the number of companies that pride themselves on being customer focused. Professor Scott Galloway points out in his book *The Four* that most of the language aimed toward customers in private industry involves a desire to distract them from policies to maximize profit. You and your utility, however, are not selling clothing or a new beauty line, but rather, you are providing an essential service that requires you to clearly address your customers who are receiving that service.

Equip Your Staff

Provide field and maintenance teams with basic information on the project duration, where the public can find information about the project, and whom to call if they have questions. Consider printing small business cards with this information that can be handed to people, if requested.

Create playbooks for customer service and articles that address common questions and implement the "warm handoff," the process of moving information and follow-up from one group or department to the next.

Practice in a group with short scenarios like a tabletop exercise (e.g., de-escalating a phone call about a high water bill, coordinating leak follow-ups, or walking a customer through a portal task). A tabletop exercise helps utility staff practice roles and responsibilities in a collaborative exercise.

MEASURE WHAT MATTERS

Throughout this book, you will see metrics and measurements that can be used by your utility to see how you are tracking success factors.

Marcus Buckingham and Ashley Goodall, in their book *Nine Lies About Work*, favor a different approach to traditional metrics, from ratings to frequent check-ins. This is true whether those metrics measure employee performance or whether they are metrics to help drive customer interaction.

There is an art and a science to defining key performance indicators (KPIs). You certainly want to use measurements to identify trends that may illustrate why customers are behaving as they are. A higher-than-average number of "high" meter reading flags may be the result of higher temperature or because of a local news story that told your entire service territory to leave the water running to keep pipes from freezing. An upward trend in unpaid bills may be due to the recent closure of a business nearby and the financial distress that comes with it. These factors can directly tie into the day-to-day performance of your utility and how you should adapt or adjust to those needs.

Use Technology to Amplify People

While this book will address the needs, data, and processes specific to advanced metering infrastructure, in general, more information does improve customer experience (within reason). Customer portals, analytics, and smart digital assistances or agent artificial intelligence tools do reduce wait time and guesswork, but too much information without helping customers understand it and guiding them to action becomes the source of more contacts and calls rather than fewer. Utilities that embrace the delicate balance for the right level of information can free staff to focus on problem-solving and empathy.

Technology should never replace judgment or care; it should make it easier to deliver both.

WHAT YOU CAN EXPECT

- Unify every touchpoint so that each one has the feel of coming from one utility.
- Use a shared tone of voice and plain-language standards for all channels, inclusive to your customer base, with accommodations compliant with the Americans with Disabilities Act as standard, not an afterthought.
- Revisit templates used for bills, notices, emails, and texts and translate explanations, steps, and timelines into everyday terms. Run them past community groups, employees, and even legislative offices to make sure they are understood by the bulk of the population.
- Work from a single knowledge base that powers customer service representatives' answers, field job aids, and self-service content.
- Develop journey checklists for common tasks (move-in/move-out, leak assistance, billing questions) to align field, billing, and communications departments. If customers can make a payment arrangement with an agent, make sure it can be done online. Help customers help themselves rather than leading them directly to contact you on the phone.
- Conduct "handoff health" audits: spot-check five cases a month to ensure customers are not having to repeat their story and the process is sticking. As Mike Tomlin, legendary football coach (formerly of the

Pittsburgh Steelers), has often noted, "The standard is the standard" (Coolong 2015).

METRICS AND RESOURCES

- Seventy percent of customers expect anyone they interact with to have the full context of their situation (Zendesk 2025).
- When customers know what a water utility is doing to maintain infrastructure and water quality, satisfaction rises significantly (JD Power 2024).
- Utilities that invest more in customer communication earn higher brand trust, especially when messaging focuses on savings programs and environmental initiatives (Escalent 2024).
- AWWA's *Water 2050* places equity, access, and community engagement as a core strategic area, embedding customer engagement in sector strategy (AWWA 2026).
- Track first-contact resolution, time to first response (known as *average speed of answer*), successfully completed payment arrangements (and after what month do they start being less successful), leak-outreach completions, and cross-department handoff success. Use monthly reviews to celebrate wins and relieve friction.

CULTURE SHOWS UP IN THE SMALL THINGS

Culture is not a poster or a slogan; it's what customers feel at 8:03 a.m. on a Monday when a pipe bursts and at 5:28 p.m. on a Friday when a bill looks wrong and they just got home from work to see it. Leaders set the tone by making service decisions visible (what they prioritize), consistent (how they act), and accountable (how they learn). Small steady choices, returning calls when promised, closing the loop after field visits, and coordinating messages between teams add up to trust.

DROPS OF KNOWLEDGE

- Treat service as a series of decisions you control, not accidents you need to explain.
- Ensure consistency, accuracy, and readability across phone, field, and digital technologies for your service territory. The brand, your reputation, is the sum of those moments of contact.
- Build culture on cross-functional standards, practice, and accountability; no single department, including customer service, owns customer trust and experiences since no single area owns all the processes that create

trust and experience. Responsibility starts with the general manager or director of the utility, not at the customer service department.

- Publish clear steps and timelines, so customers know what will happen and when.
- Use data and tools to personalize help rather than deflect help.

Responsibility is the starting point. Culture is how responsibility shows up.

In the next chapter, we move from the idea that service belongs to everyone to what that looks like in practice and how a unified voice, consistent messaging, and visible follow-through build trust, even when customers have no alternative provider.

WORKS CITED AND BIBLIOGRAPHY

AWWA. 2026. *Water 2050.* www.awwa.org/water-2050/ (accessed February 26, 2026).

Buckingham, M., and A. Goodall. 2019. *Nine Lies About Work: A Freethinking Leader's Guide to the Real World.* Cambridge: Harvard Business Review Press.

Coolong, N. 2015,"'The Standard is the Standard' explained." *Steelers Wire.* https://steelerswire.usatoday.com/story/sports/nfl/steelers/2015/10/28/mike-tomlin-tomlinisms-the-standard-is-the-standard-explained/81036494007/ (accessed February 26, 2026).

Escalent. 2024. *Utility Trusted Brand & Customer Engagement Study.* Livonia, MI: Escalent.

Galloway, S. 2017. *The Four: The Hidden DNA of Amazon, Apple, Facebook, and Google.* New York: Portfolio.

JD Power. 2024. *U.S. Water Utility Residential Customer Satisfaction Study.* Troy, MI: JD Power.

Schloeffel, J., and C. Firth. 2025. *Wankernomics: A Deep-Dive into Workplace Bullshittery.* London: John Murray Press.

Walton, B. 2022. "Tap Water Failures and Distrust of Government: A Conversation with Manny Teodoro." *Circle of Blue.* www.circleofblue.org/2022/supply/water-management/infrastructure/tap-water-failures-and-distrust-of-government-a-conversation-with-manny-teodoro/ (accessed February 26, 2026).

Zendesk. 2025. *Customer Experience Trends 2023–2025.* San Francisco: Zendesk.

Chapter 2

Earning Trust Without Choice

Even when customers cannot choose their water provider, they still judge the experience. They compare it to their bank, their internet company, and the best service interaction they have had anywhere. In a monopoly environment, trust is not earned through competition; it is earned through consistency.

If Chapter 1 established that service is everyone's responsibility, this chapter focuses on what that responsibility looks like in action. Culture becomes visible when customers hear the same message from every role through every channel, especially when pressure is high.

Some in the water industry ask: *As long as the water is clean and reliable, isn't that the main focus?*

The short answer: *No.*

Clean and reliable water is the baseline. Trust is built in the moments when customers have questions, concerns, or confusion. Every interaction—from a billing dispute to a taste and odor question—becomes a public display of your utility's values.

Most utilities include a commitment to service in their mission statement or strategic plan. Whether it's phrased as "meeting the needs of stakeholders" or "delivering exceptional service," those words carry weight only when they appear in daily practice, visible in how staff listen, explain, and resolve. Culture is not what is written. Culture is what is repeated.

A UNIFIED VOICE: EVERY ROLE, EVERY CHANNEL

Field crews, plant operators, call center staff, billing specialists, and public information officers are all part of one customer voice. When a resident stops a field technician on the street to ask why the water is shut off and for how long, the answer they receive should match what they hear on the phone or see online.

You do not need every detail to reassure a customer. A confident consistent message about when the water may be restored builds trust. Inconsistent messaging erodes it.

Research supports this connection between communication and trust. Customers who believe their utility maintains infrastructure well report 229 points greater satisfaction on a 1,000-point scale (JD Power 2024).

STAY IN SYNC

Technology engagement tools such as internal messaging platforms help teams coordinate during outages or water quality events. Coordinated updates between field, dispatch, customer care, and public information employees prevent the frustrating "handoff" feeling that customers dislike and help the utility speak with one voice.

Satisfaction is built across eight dimensions (JD Power 2025):

- Trust
- Quality and reliability
- Ease of doing business
- Information provided
- People
- Problem resolution
- Digital channels
- Monthly cost

Trust also depends on who delivers the message. Seventy percent of US adults trust their water utility "a lot" or "some" for information about tap water (AWWA and Morning Consult 2024).

That level of trust is the true gem in this industry, but it must be reinforced through consistent behavior.

A UTILITY TALE

This story was told by one utility.

A customer calls to say their water has an off taste. It's midsummer, and the likely cause is an algae-related taste and odor compound in the source water. You could brush it off because it is not posing a safety risk: "It's safe; everyone has it."

But you don't.

This is a "teachable moment," as President Obama once said, giving you an opportunity to reinforce trust, not because of one contact or transaction but because you are the source of care.

We spend time explaining the situation in plain language, answering questions, offering tips (like chilling water or using a filter), and reassuring customers that their concern matters and that the water does not pose a threat to their family. That single interaction reinforces not just the safety of the water but the credibility and care of the people behind it.

DROPS OF KNOWLEDGE

- Make the mission visible: Translate values into scripts, service standards, and follow-through.
- Speak as one utility: Align field, plant, and customer care messaging, especially during incidents.
- Close the loop: Acknowledge the concern, explain the why, set expectations, and follow up.
- Use data to coach: Track first-contact resolution, callbacks, and complaint themes that need improvement.

When considering how to approach these efforts, think through people and the data. Data should not only measure performance; it should reinforce culture.

Next, we connect how the execution of customer service is strongly tied to the very nature of your customer persona, whether an internal or external stakeholder, employee, business, or residential customer.

WORKS CITED

AWWA and Morning Consult. 2024. *Public Perceptions of Tap Water Survey.* www.awwa.org/wp-content/uploads/Public-Perceptions-of-Tap-Water-Survey-24-Slides.pdf (accessed February 26, 2026).

JD Power. 2024. *U.S. Water Utility Residential Customer Satisfaction Study.* Troy, MI: JD Power.

JD Power. 2025. *U.S. Water Utility Residential Customer Satisfaction Study: Methodology Overview.* Troy, MI: JD Power.

Chapter 3

Who Are Your Customers?

Before we go any further into how to deliver outstanding service, let's be clear about the community receiving your service. In your water or wastewater utility, customers are more than the people who remit a monthly payment. They include every person, group, or organization that relies on your work or is affected by it every single day.

EXTERNAL CUSTOMERS

Let's start with the customer receiving the bill: the residential, commercial, and industrial accounts that receive the water your company treats and distributes (and, on the wastewater side, the collection and treatment services you provide). These customers experience your service when they

- turn on a tap or flush a toilet;
- call the utility, email, or chat online to ask a question;
- visit a payment center or your city hall to start, stop, or transfer service;
- request field support for a leak, quality concern, or meter issue; or
- look at the water in their drinking glass.

These individuals and organizations are best characterized as external customers because they exist outside of the staff organization whose job it is to provide both the technical and personal service of water delivery. For these customers, water is personal; they judge a utility not only on flow, pressure, and taste but on every interaction that surrounds those essentials.

Internal Customers

By comparison, if not by rule, everyone on the utility team that serves the external customer is an internal customer. If we begin from the perspective of a water professional, the first set of internal customers are your fellow team members—the hardworking, reliable, professional voices and faces of the utility.

Other water professionals (i.e., your coworkers) are internal customers as well:

- Supervisors and managers, who set priorities and remove obstacles

- Public information officers, who develop messaging shared with the public and customers, help keep communication consistent across all platforms, and build customer confidence through uniform messages
- Field services crews like meter readers, service techs, and construction teams
- Treatment plant and laboratory staff, who safeguard water quality
- Human resources, finance, IT, and every shared service colleague

Everyone who works under the same utility logo that you'll find on company vehicles, uniforms, letterhead, or staff paychecks is an internal customer counting on you. Great internal service starts with competent, positive, professional execution of utility leadership's own duties and grows into mutual respect, shared commitment, and a sense of teamwork.

The Golden Rule relates to customer service: Behavior that would be recognized as great service to a customer includes listening actively, responding promptly, and solving a problem. The same behavior should be extended to a coworker who needs assistance or raises a sensitive issue in a staff meeting. Show up every day and serve everyone well.

Stakeholders

A special class of customers we call *stakeholders* may or may not pay a water bill, yet they have a vested interest in how the utility operates because utility performance touches their organization, livelihood, or quality of life. What follows is a list of common stakeholder groups.

- Governing body: the board, commission, or council that provides budget oversight, policy direction, and strategic vision
- Regulators: state primacy agencies (or for Wyoming, the US Environmental Protection Agency's Region 8) and local health departments that enforce drinking water and wastewater regulations
- Public works partners: departments responsible for streets, traffic, bridges, and public transportation that share rights-of-way with utility pipelines
- County commissions: If your utility is billing water and the county, a separate agency, is billing sewer service
- Public safety agencies: police, fire, and emergency medical services staff who depend on reliable flows, pressure, and hydrant performance
- Other utilities: power, gas, broadband, and waste management providers whose projects and outages intersect with a water utility
- Parks and recreation departments: managers of parks, golf courses, and sports fields that rely on your water resources
- Business community: chambers of commerce, developers, and major employers that drive economic growth
- Environmental and water advocacy organizations: watchdogs and allies in resource stewardship
- Schools and community groups: beneficiaries of a utility's education and outreach programs

There are more, but you get the point: Your external customers and stakeholders are fundamentally why your utility exists, to meet water and wastewater needs 24/7/365. Cultivating these relationships is both prudent and essential.

FROM CONCEPT TO CULTURE

A utility needs to know who its customers are, but that's not enough. How do we really "know" a customer? A true service culture emerges when every employee

- recognizes the customer or stakeholder in front of them (on the phone, in person, or across the organization);
- adapts his/her communication style and problem-solving approach to the situation at hand; and
- acts with the Golden Rule in mind, delivering the experience they would want if the roles were reversed.

In the book *Peak Mind* by Dr. Amishi Jha, there are some fundamental skills that should be cultivated to provide these experiences to customers. Says Jha, "Your attention is like a flashlight; it can only illuminate one area at a time. If it's scattered, you miss things. To know someone, you must shine your focused attention on them, not on your internal chatter" (Jha 2021).

In addition, knowing that our brains wander about 47% of the time, planning and staffing are critical to perform the type of engagement needed to provide good customer service.

Ingraining these behaviors into policies, training, daily work habits, and how the utility engages with all customers turns good intentions into consistent performance.

A UTILITY TALE

A medium utility in Colorado recognized the needs of its growing, new, younger customer group and committed to doing more to connect with them. That effort involved tracking and reporting the utility's stakeholder outreach efforts.

Utility staff, when interviewed, identified the behaviors they now wanted to incorporate into their operations. In some examples, employees resolved to

- contact all affected customers within 30 minutes of any water main issue,
- respond within the first 24 hours of any inbound email or correspondence, even if the response is that additional time may be needed, and
- provide a debrief on more substantial customer service issues after they happen, listing positive and negative elements as takeaways.

For each item, utility staff gave themselves a score, with 0 = not doing it, 1 = doing it at least once in the past year, and 2 = doing it regularly (if not annually). In the first year of tracking and scoring, the utility scored 12 out of a possible 16 points and will strive to improve from one year to the next.

How do your typical activities rate: communicating with your customers, receiving feedback, and incorporating feedback into your practices?

DROPS OF KNOWLEDGE

- Have one standard: Treat internal and external customers with equal respect and diligence.
- Stakeholders matter. Many groups depend on utility performance, even if they never receive a bill.
- Service is everyone's job. Every role, from lab analyst to backhoe operator, influences the customer experience.
- Paying attention is a skill that can be improved and refined with practice, and it is critical to customers' feeling heard. Work environments, training, and even workspaces should be nondistracting to create the atmosphere needed for attention.

In the next chapter, we will focus on the traits of customer service. Now that you know who your internal and external stakeholders are, let's analyze the context for what you do, how you do it, and why.

WORK CITED

Jha, A. 2021. *Peak Mind: Find Your Focus, Own Your Attention, Invest 12 Minutes a Day.* San Francisco: HarperOne.

Chapter 4

The Magic of the Customer Experience

The magic of the customer experience is that it is just part of your day-to-day role. It is not tucked away in a group or in a department but is rather a core requirement of every job in clean water. Realistically, however, it may not be that simple. The definition of customer service varies as much as the utilities. San Jose Water provides this definition: "For us, customer service means being available to answer your questions promptly, respond quickly to emergencies, provide a safe and reliable water supply, and much more."

This message is not just culture but a standard of professionalism and way of operating. It's a standard of professionalism that should guide the entire utility, whether staff are in the field, managing finances, or monitoring operations behind the scenes.

TRAITS OF A PROFESSIONAL

Traits, not job titles, describe true customer service professionals. These are the everyday attributes that define exceptional service for any position in a utility.

- Recognizes and works to meet customers' needs
- Always performs to the best of their abilities
- Keeps knowledge and training up to date
- Maintains a positive attitude that customers can sense, even over the phone
- Delivers what is promised and avoids overpromising
- Communicates clearly and in a way that customers can understand
- Acts ethically and honestly
- Shows empathy, especially during hard conversations like service disconnection or complaints
- Follows up to ensure an issue has been resolved
- Remains calm, even in stressful situations

THE PSYCHOLOGY OF CUSTOMER SERVICE

Working in customer service can be both rewarding and stressful. Speaking with people, solving problems, and educating customers often bring relief and clarity, but they can also bring frustration, particularly when tensions are running high.

Someone who has not worked in a utility's customer-facing role may have heard employees giving replies such as these:

- *"Lower your voice; you don't need to scream at me."*
- *"I'm explaining; don't talk over me."*
- *"I didn't set the rates; they're available on our website."*
- *"I can't set up a payment plan; you've already broken five."*
- *"We'll fix the grass after the repair."*

Or maybe they overheard one employee telling another, *"Customers never pay their bills and still blame us when the water is shut off."*

These aren't easy words to hear. Working in customer service directly often means wearing different hats—part educator, part therapist—all while keeping up with daily operations. And now, with the pervasive role of technology, it's all the more important to understand the human psychology behind customer interactions.

A DIFFERENT KIND OF MARKETPLACE

Unlike traditional businesses, most utilities operate without competitors. Rather than having the option to shop around, water customers inherit a water utility depending on where they live or work. Their "purchase" moment happens when they turn on the tap. A utility delivers the service, meters the usage, and sends a bill. If all goes well, customers barely notice. But when the bill arrives, the utility enters the spotlight: *"Does this look right?"* the customer wonders. If the numbers seem off, questions will come up:

- *Was there a leak?*
- *How could I have used all that water?*
- *Did the sprinklers run too long?*
- *Did we have houseguests?*
- *Is this meter reading accurate?*

The anonymous water provider becomes real, and the customer wants answers. Or what about the new resident from out of state, the first-time homeowner, the new business owner hooking up service, or the builder developing a new neighborhood? Their expectations and experiences may differ, but their need for clarity and connection remains the same. That's when good customer service matters most.

If there are customers, there will be questions, issues, and discrepancies, so there must be answers and solutions. How effectively those are delivered and how professionally and empathetically you communicate them shapes how customers perceive you and your utility.

J.D. Power measures water utility customer satisfaction across key dimensions that extend beyond price alone (JD Power 2023):

- Quality and reliability
- Conservation
- Problem resolution (billing and payment)
- Ease of doing business (customer service)
- Digital channels (communications)
- People (customer service)
- Monthly cost (price)
- Information provided (communications)

METRICS AND RESOURCES

Customer experience is not accidental. It is behavioral.

- Customers who rate utility employees highly for professionalism and courtesy report significantly higher overall satisfaction scores (JD Power 2023).
- According to the AWWA *State of the Water Industry Report*, utilities increasingly identify workforce development and customer trust as top strategic priorities, recognizing that service outcomes depend on trained, engaged staff (AWWA 2024).
- Research from Gallup shows that employees who feel engaged in their work are 21% more productive and deliver stronger customer outcomes than disengaged employees (Gallup 2023).
- Studies in service psychology about the fair process effect have demonstrated that customers are more likely to accept unfavorable outcomes like rate increases or disconnection policies when they believe they were treated fairly and respectfully during the interaction.

Professional traits are not "soft skills." They are measurable drivers of trust.

DROPS OF KNOWLEDGE

- Customer service is everyone's job. Even if you never speak to a customer, your work still affects their experience.
- Every interaction is a reflection of your utility. A calm, clear, professional approach makes a lasting impression.

- People remember how you made them feel. Your approach determines whether they leave frustrated or reassured.

No matter your title at a water utility, your work shapes the customer's experience. Next, we'll put this mindset into practice, introducing communication tools to make your message clear.

WORKS CITED

AWWA. 2024. *State of the Water Industry Report.* www.awwa.org/state-of-the-water-industry/ (accessed February 26, 2026).

Gallup. 2023. *State of the Global Workplace 2023 Report.* Washington: Gallup.

JD Power. 2023. *U.S. Water Utility Residential Customer Satisfaction Study.* Troy, MI: JD Power.

Chapter 5

Communication Skills, Tools, and Techniques

In the previous chapters, we discussed the traits and roles that shape strong customer service. In this chapter, we'll explore the skill that brings those traits to life: communication. As a clean water professional, you are the voice of the utility. Whether you're responding to a billing question, walking someone through a service startup, or listening to a complaint, the way you communicate shapes how your customers feel about your organization.

As mentioned in Chapter 3, even internal communication with coworkers or external stakeholders who do not receive service from you can ripple outward to affect how service is delivered. Great communication doesn't happen by accident. It's a practiced skill that requires time, feedback, and awareness. Let's walk through some foundational communication concepts, along with practical tools to help you build and refine them.

The utility world may not look like corporate America, but the fundamentals still apply. Great leaders and great philosophers agree on one thing: How you communicate matters just as much as what you say.

> "We have two ears and one mouth so that we can listen twice as much as we speak."—Epictetus, ancient stoic philosopher

> "In every interaction, you have to account for your intention, your words, and how the other person hears you."—Deborah Liu, former CEO of Ancestry.com

> "The simple act of talking about what goes into delivering safe and affordable water can have a big impact on how people perceive water quality."—Cheryl Porter, past president of AWWA

> As Peter Parker's Uncle Ben said in *Spider-Man* (2002), "Remember: with great power comes great responsibility."

Starting with the basics will help you effectively use your responsibility as a communicator.

THE COMMUNICATION PROCESS

At its core, effective communication has four components:

- Sender: the person delivering the message
- Message: the actual content of the communication
- Receiver: the person receiving the message
- Feedback: the response, which tells the sender how the message was received

When communication is working well, the message sent is understood as intended, and the receiver responds in a way that confirms understanding (Figure 5-1).

What is often lacking from discussions about that communication process are the skills that are required in order to execute that process successfully. The effectiveness of individuals being adaptable, having interdisciplinary knowledge, and feeling confident with changing technology is core to making the communication process work (Chang 2024).

Since the process of communication does not always succeed, an examination of some of those root causes can help improve the function.

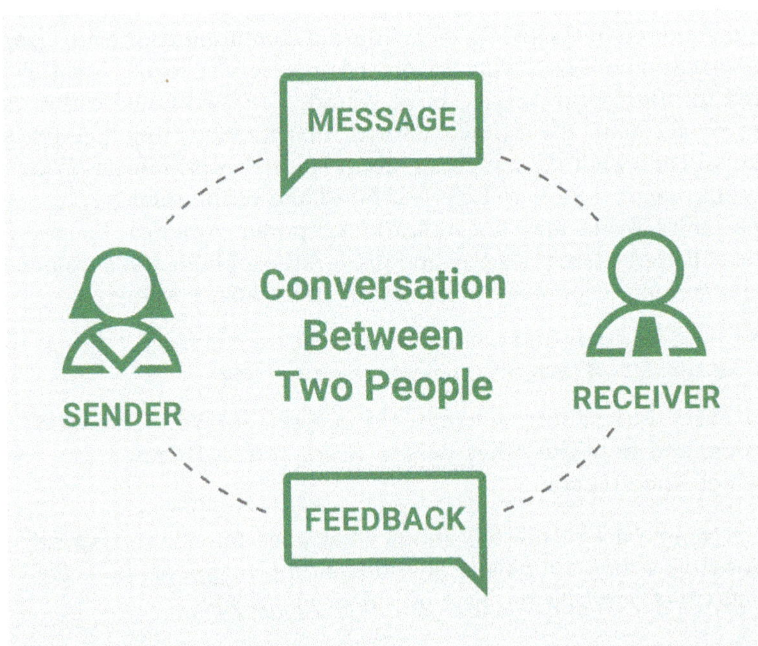

Source: AWWA 2026

Figure 5-1 Exchange of communication flow

Skills That Support Positive Customer Engagement

- Listening: Perhaps the most underrated skill, active, engaged listening shows respect and improves accuracy.
- Presenting: If visuals such as bills, maps, or service notices are part of your message, make certain they are easy to read and understand for your service territory.
- Speaking: Speech should be in terms that reflect how a customer or stakeholder must understand the message. Individuals who work with customers do not have to speak overly formally, since matching someone's speech may often bring comfort and familiarity, but it must be understood by the listener. A steady pace, appropriate volume, and avoiding acronyms or terms that may be familiar to your utility but not to the customer are markers of appropriate speech. ("I'm looking at your AMI, and I'm seeing some spikes" is a common example of jargon usage that leads to an unhappy experience for the customer.)
- Questioning: Use questions to clarify, not to confront. This shows your interest in understanding the customer.
- Evaluating: Assess both what is said and how it's said. In the case that this communication is in the field, at the customer's house, or face-to-face in the office, look for visual cues of underlying concerns or confusion.
- Observing: Body language and tone often reveal more than words.
- Positive attitude: Approach every interaction with sincerity and an open mind. A helpful tone goes a long way.
- Assume good intent: When both parties start with that premise, it is much easier to approach difficult conversations objectively rather than in a dehumanized way.

Understanding Communication Styles

It goes without saying that knowing your style and being aware of others' ways of communicating drastically improve the outcome. Adjust to match what the customer needs: e.g., bottom line versus details, moving quickly versus moving slowly, striking a very formal tone versus a more conversational or lightly informal style.

Keep in mind that a customer service representative who takes a direct approach may need to slow down for a customer with a more deliberate manner or processing speed time, and a talkative representative should keep answers focused for a customer who likes to get right to the point.

Levels of Listening

Consider these six levels of listening, knowing that between 30–50% of our waking time is spent with our minds wandering (Jha 2022), from least (1) to most (6) effective:

1. Ignoring: tuning out entirely
2. Pretending: feigning attention while being mentally elsewhere
3. Monologuing: like an actor in a theater, listening solely for the purpose of speaking
4. Being selective: hearing and retaining only what supports your view
5. Being attentive: focused and engaged
6. Being empathetic: fully present, seeking to understand

We all shift between these levels, especially under stress. But to deliver great customer service, aim for level 5 (attentive) or level 6 (empathetic) as often as possible.

Steps to Better Listening

Listening is one of the most important aspects of effective communication. The film *Recovering the Stories: Exploring the History and Resilience in Dallas Communities* (Hawkins 2023) emphasizes how listening underpins many experiences, from community development to utility customer service. Follow these tips to help ensure a positive customer experience.

- Listen to understand, not just to respond; to help, you will need a complete and accurate history before you can try to assist.
- Be quiet and let the customer talk. To feel heard, a customer must have an innate sense of psychological safety to be open about financial or housing issues.
- Don't interrupt, even if you're excited to help.
- Ask questions to confirm your understanding.
- Use phrases like "Let me make sure I heard you right." Validation is a big part of building trust within the utility–customer relationship. Your utility also bears the entire history that an individual has had with public entities, so discounting that background invalidates a customer's known background.

Tone, Volume, Pace

Consider three mechanisms that influence how your message is received.

- Tone: Are you patient, or are you rushed? Are you encouraging, or are you dismissive? Smile when you speak; it shows, even over the phone.
- Volume: Focus on being not too loud or too soft. Adjust your speaking volume according to the surrounding environment, the customer's age, the customer's hearing ability, and the situation prompting the conversation.
- Pace: Match the customer's pace. Slow down if needed. If you speak quickly, you may need to pause to let the customer catch up.

A UTILITY TALE

One utility shared a tale illustrating the importance of listening. A customer service representative (CSR) was speaking to a customer about their water bill. The customer was confused about the amount due and the reason for the amount.

The CSR tried to explain to the customer that they were paying only part of the bill each month and therefore carrying a balance forward. As a result, the customer incurred late fees and built up a large past-due amount. The customer did not understand the explanation and got frustrated. The CSR acknowledged and reassured the customer that a bill can be confusing and reaffirmed that the utility wanted to help.

Deciding to try a different approach, the CSR downloaded the bills and emailed them directly to the customer, then reviewed one bill at a time, so the customer could visually see what caused the balance due. The customer then understood and was satisfied.

Thinking about what took this customer from frustration to satisfaction shows the importance of great customer service.

- Acknowledgment: The customer did not feel alone or that they were talking to a faceless person across the phone line.
- Reassurance: Empathy helped to defuse the situation.
- Adapting: The CSR was ready to adapt the approach to suit the situation.
- Action: Instead of asking the customer, who might not be set up to download the bills online, the CSR made it easy to see the same information.
- Empowerment: The CSR was empowered to try something new rather than hurrying to finish the call.

Another common example follows from another utility, this one dealing with water-leak adjustments.

A customer service supervisor dealt with an escalating conversation with a customer who thought that, rather than receiving a partial credit, she should get full credit for all the water that was lost during a leak event. The supervisor explained the leak adjustment policy, but the customer was still upset.

The supervisor listened and empathized with the customer and then took a different tack, explaining that the policy was designed as a shared responsibility between the utility and the customer. As a community-owned utility, fairness needed to be extended to all customers. Although the customer still wanted a larger credit, the customer understood it was not possible and was satisfied. Using the same skills as the CSR did in the first example, this supervisor took a difficult situation and resolved it successfully.

METRICS AND RESOURCES

- Clear, proactive updates correlate with higher satisfaction; utilities that leverage digital channels effectively see measurable gains in customer sentiment. As your utility may be at various points in the artificial intelligence journey, these same criteria of listening should be trained into the operations of any digital engagement just as consistently as their people.
- Utilities that score higher in the "People" and "Information Provided" dimensions report stronger overall satisfaction (JD Power 2025).
- Invest in understanding your communication style and/or that of your employees. This should include more standardized methods of assessment, such as the Myers-Briggs Type Indicator, but also practical methods like peer-to-peer practice, shadowing, and input from independent third parties.
- According to AWWA's *State of the Water Industry Report*, workforce development and customer trust remain top sector priorities, reinforcing that communication is a strategic asset, not a soft skill.

DROPS OF KNOWLEDGE

- Communication is central to customer service. It is how you build trust, explain policy, and solve problems.
- Listening intently/actively matters as much as speaking clearly. People want to be heard before they want to be helped.
- Adjusting your tone, volume, and pace of speech improves connection. Minor changes can make a big difference.

Keep practicing. You do not need to be perfect to be professional. Every conversation is a chance to improve your skills and help someone feel heard and respected. Asking peers and listening to them will help each CSR develop a professional communication style.

Next: How to handle difficult customers with calmness, confidence, and care.

WORKS CITED

AWWA. 2026. *Customer Service Training Program. Course 1: Customer Relationship Building.* Denver: AWWA.

Chang, H. 2024. "Strategic Directions for Business Communication Education in Response to AI-Driven Industry Changes." *Business Communication Research and Practice,* 7:2:71–73.

Hawkins, J. 2023. *Recovering the Stories: Exploring the History and Resilience in Dallas Communities.* Arlington, VA: Public Broadcasting Service.
JD Power. 2025. *U.S. Water Utility Residential Customer Satisfaction Study.* Troy, MI: JD Power.
Jha, A. 2021. *Peak Mind: Find Your Focus, Own Your Attention, Invest 12 Minutes a Day.* San Francisco: HarperOne.

Chapter 6

Handling Complaints and Dealing With Challenging Situations

Working in customer service for a water or wastewater utility is different from working in retail, food service, or hospitality. At a restaurant, customers choose to walk through the door. This gives people a sense of freedom of choice. In the water industry, however, a utility may be the only one for miles, and depending on where you live, there may be no choice for customers but to get their water from that utility (if they aren't able to dig their own well). Still, that should not change customers' expectations about the level of professionalism, care, and problem resolution they receive.

In addition, in many cases, water customers may have little or no understanding of the technical side of the business. I don't know how many of us really remember the water cycle from middle school science class. You are the bridge between the utility and the customer's experience. How you handle complaints and difficult situations is just as important as how the utility treats water before it reaches the tap.

UNDERSTANDING THE UPSET CUSTOMER

A common challenge: A customer is upset when the water is turned off as the result of a leak. Their frustration may be justified—something went wrong, and your job is to restore water service. The opportunity: to build a customer's trust.

This is where service recovery comes in. These five steps can help rebuild trust and restore service quality:

1. Apologize and sincerely acknowledge the issue. Just as in addressing concerns about high bills, the customer service department represents the entire utility. Acknowledging an issue, even if it has nothing to do with your specific job, is just as important.
2. Find out what it takes to fix the problem now. Act swiftly to determine the remedy.

3. Offer a fair and reasonable fix. Be clear and realistic about what you can deliver. If it takes half a day to fix a system leak, prepare the customer. If you have no control over the timing, say that you cannot yet provide a time for when the issue will be resolved and that you will follow up once you have more information.
4. Follow through. Deliver precisely what you promised.
5. Follow up. Confirm that the problem is resolved and that the customer is satisfied.

That final step is critical. Research from SQM Group shows that for every 1% improvement in first-call resolution (FCR), customer satisfaction increases by 1% and satisfaction drops by 15% when a customer has to call back (SQM Group 2023).

TURNING DOWN THE HEAT WITH "HEAT"

An additional approach you can use to de-escalate customer conflicts is the HEAT method:

- **H**ear them out. Let them finish speaking, without interruption.
- **E**mpathize. Show that you understand their frustration.
- **A**pologize. Even if you're not at fault, you can apologize for the inconvenience.
- **T**ake responsibility. Own the next steps toward resolution.

Managing Conflict Step-by-Step

Professionalism and clarity typically are enough to successfully handle customer issues. Occasionally, though, challenging calls can lead to a higher degree of tension. When there's a conflict, start by determining exactly what the problem is so that everyone is on the same page. Unless you both agree on the nature of problem, you will not find a solution.

This is an example of how you might respond to a customer's call about a high bill: "If I understand you right, you believe that this month's bill is incorrect? Or are you disputing the past-due amount on the bill that is shown?"

Then, look for possible solutions that could work. Try to find one that both sides can agree on, ensuring that the customer feels heard and that the utility can meet its responsibilities.

> "Do you want to walk through the bill together? We can figure out where the balance came from and how the bill was calculated."

Keep emotions in check, and don't let anger get in the way of solving the problem. The goal is to fix the issue in a way that works now and also prevents the same problem from arising again.

> "I understand that you are upset because you don't think you used that much water, but it can be hard to judge the amount of water we use

every day. I can ask you some more questions about your household to give you estimates that you can use going forward. And while I do not advise it, you can perform a meter test for a charge that we will refund if your meter is running too fast. Bear in mind that it's unlikely that this will be the case. You can also sign up for our online service so that you can monitor your water use." (Note that unless your utility has meters that use advanced metering infrastructure to transmit hourly data, the last suggestion may not be relevant.)

In this example, the customer has options: either going into further detail on the bill or deciding to test the meter (understanding that it's not likely to change the outcome). In addition, the customer has options for the future, such as being able to monitor their usage more frequently online and making adjustments to their lifestyle to conserve water. While this approach may not produce a happier customer, it can defuse the situation by keeping the focus on control and options.

Effective communication can have a significant effect: The 2024 US Water Utility Residential Customer Satisfaction Study found that when customers believe their utility communicates effectively about infrastructure work, satisfaction increases by 229 points on a 1,000-point scale (JD Power 2024).

WORKING WITH CHALLENGING CUSTOMERS

When you're dealing with an upset customer, expect strong emotions. Don't take their complaints personally; most of the time, their frustration is aimed at the situation, not you. Give them space to share their story and listen closely for details that will help you uncover the real issue.

Once you know the problem, work quickly to find a resolution. Acting promptly keeps the situation from growing larger and shows the customer you take their concerns seriously. If the customer crosses the line into insults or threats, remain calm, but be clear that such behavior is unacceptable.

After an issue is resolved, follow up to confirm the customer is aware of the outcome and, if possible, satisfied with it. Thank them for bringing the matter to your attention; this reinforces the idea that you value their input. Maintaining a positive, composed attitude through these steps improves communication, strengthens teamwork, builds staff morale, and helps keep productivity high.

Handling Customers Who React Abusively

Some customers may become verbally abusive, and while anger directed at someone who has no control over a situation is never acceptable, keep in mind that people usually display this response when responding to fear. When it comes to a necessary service like water, the idea of potentially losing water service because of the inability to pay a high bill, for example, can lead to unpredictable behavior. Remain professional and know your policy.

Stay calm and clearly state boundaries: "I must end this call if the abusive language continues."

De-escalate the situation if possible: "Let's talk through this issue. I can understand you are upset." Finally, if you feel unsafe, involve your supervisor or security (if a threat is made).

Supervisors and managers should step in to assist, and when customer hears the same message from another person in the utility, there is an opportunity for you to feel more confident and the situation to de-escalate.

Accepting That Some Customers Cannot Be Satisfied

Some people may always be dissatisfied, regardless of your efforts. If customers' household budgets are stretched, complaints can rise with the financial stress. Focus on maintaining your standards. If you've behaved professionally, your supervisor will support you, even if a complaint follows.

To put it in context, at one water utility location during 2023–2024, unresolved utility complaints rose by 29% to 222,956. Of those complaints, 57% were about billing. A substantial change in economic factors had followed a massive job loss within the service territory.

METRICS AND RESOURCES

- For every 1% improvement in first-contact resolution, customer satisfaction increases by approximately 1%. Repeated contacts drive dissatisfaction and erode trust (SQM Group 2023).
- Industry guidance targets approximately 70–75% FCR for core live channels; improving FCR also reduces operating costs.
- Tracking complaint themes and time-to-closure reveals systemic issues that can prevent repeat calls and reduce escalation.

Measures:

- FCR
- Repeat contact rate
- Complaint category trends
- Time to resolution
- Escalation frequency

A Note on Artificial Intelligence (AI) and Digital Assistants

While this book does primarily address human-centered work, preparing the team for the influx of digital tools to address difficult or conflict situations is just as important as training people on the task. Using senior staff who model the de-escalation and resolution process well to train digital agents will aid in preparing your workforce for a different type of work, while reemphasizing the need for humans in the technology mix. And while AI continues to change the working mix, conflict and difficult situations in public service often still revert to human-to-human contact.

DROPS OF KNOWLEDGE

- Service recovery is about restoring trust, not just fixing problems.
- Think just as critically about the path of resolution and communication for your digital assistance as you do your people.
- Listening, empathy, and clear communication are essential tools, especially in tense moments.
- Proactive, timely follow-up reduces frustration and increases customer satisfaction.
- Even when some customers can't be fully satisfied, professionalism must remain constant.

In the utility field, you are often meeting customers at their most frustrated: when their bill is too high, the water is shut off, or they feel unheard. Your ability to de-escalate, restore, and follow through is more important than any social media post. How you handle today's issue builds tomorrow's trust. When showing up for your customers becomes routine, give your water a name and a voice so every touchpoint sounds like you rather than a generic utility.

Next: branding, that ever-essential element to make it uniquely yours and your customers.

WORKS CITED

JD Power. 2024. *U.S. Water Utility Residential Customer Satisfaction Study.* Troy, MI: JD Power.

SQM Group. 2023. *Contact Center Customer Service QA State of the Industry Benchmark Report.* Coeur d'Alene, ID: SQM Group.

Chapter 7

Your Water = Your Brand

Note: This chapter draws on the work and public presentations of Kelley Dearing Smith, Vice President of Communications & Marketing at Louisville Water, and references the "Louisville Pure Tap®" program.

Utilities communicate. Brands connect. Branding is more than a mark or a tagline: It's the story, personality, and behaviors that earn trust over time. Kelley Dearing Smith, the vice president of communications and marketing from Louisville Water (Louisville, Ky.), speaks extensively on giving tap water an identity customers recognize, request, and recommend.

BRANDING

Branding goes beyond project notices and technical updates. It's the human-centered story of who you are, what you stand for, and how you consistently show up. Louisville Water describes branding as the identity and voice people remember when you're not in the room; done well, it creates emotional attachment and trust.

Louisville Water and the team overseeing communications and marketing practice brand journalism: a blend of journalistic discipline and content marketing that focuses on relationships, third-party endorsements, and community engagement delivered across platforms. They operate from research, routine tracking of satisfaction, quality perceptions, and brand reputation; then they apply those findings to word choices and visuals.

Four principles guide Louisville Water's approach.

- Always start with people and keep the core promise.
- Promote high-quality, reliable water, keeping it front and center.
- Promote what's unique about your utility and its history.
- Think strategically about your message—your story—and how and where you'll share it and how you'll know if it's received.

National surveys indicate strong satisfaction with tap water and high trust in utilities and scientists for information. The US Environmental Protection Agency's (USEPA's) Consumer Confidence Report (CCR) rule that water utilities' annual reports are due to customers by July 1 gives utilities a built-in moment to reinforce brand clarity and trust with plain-language quality results.

A UTILITY TALE

Louisville Water's "Louisville Pure Tap"

Louisville Water has talked about its water quality and reliability for more than a century, but a pivotal step came in 1997 when it registered the trademark Louisville Pure Tap®. The idea sprang from repeated public praise for the taste of its water. The utility treats aesthetics (taste/appearance) as early indicators of perceived quality, integrating taste tests into routine parameters. To spark a conversation about water, Louisville Water trademarked its water to create a platform to talk about the quality, innovation, and value of tap water.

Branding starts with a purpose and, in this case, the product as well. Louisville Water puts on-brand service into action by serving Pure Tap in pitchers and compostable cups; equipping employees with reusable bottles; and featuring mobile units, coolers, a "water bar," and a network of bottle-filling stations. In 2010, the utility attached the brand to the clearest signal of available drinking water: the filling station. Now there are more than 100 stations across the community, with messages tailored according to location (e.g., hydration at sports venues, "locals love the water" messages at airports and convention centers, education on campuses).

The team treats locations as partners, often with formal marketing agreements and expectations (e.g., allowing reusable bottles). Dedicated staff manage partnerships, events, education, and outreach with storytellers who emphasize narrative and social media over traditional one-way communications.

Branding also fuels economic development. Distilleries and food-and-beverage businesses use Louisville Water in their recipes, and because water is bourbon's "best friend," the Pure Tap identity appears in tastings, visitor experiences, and recruitment trips. The city now lists water as a leading asset in its economic plan.

Consistency pays off. In recent research, nine in ten customers said they trust Louisville Water to provide safe drinking water; brand recognition for Louisville Pure Tap is similarly strong. Customers frequently describe the utility as reliable, friendly, respectable, trustworthy, and honest.

METRICS AND RESOURCES

- Eighty-seven percent brand recognition: customers identify Louisville Pure Tap as Louisville Water's product.
- Words matter: Swap jargon for plain language and lead with what customers care about (service, safety, taste, reliability).
- Bottled-water backdrop: Per-capita consumption continues to rise nationally, an external market force your local brand must counter.
- Use graphics and colors to highlight your data from your CCR to any document sent out to your population. The KISS ("keep it simple, stupid") principle always works.

DROPS OF KNOWLEDGE

- Name the product. A clear, consistent product identity (e.g., Louisville Pure Tap) anchors the brand to what you deliver every day.
- Make the product visible. Put your brand where the water is (bottle-filling stations, mobile units, water bars), and tailor messages to each setting.
- Build a storyteller team. Treat communications as brand journalism; invest in research and people who turn data into narratives that customers care about.
- Choose words with care. Avoid jargon and abbreviations (especially in headlines) on topics like per- and polyfluoroalkyl substances (commonly known as PFAS) and lead service lines or lead in water; communicate with relevance.
- Partner with purpose. Formalize co-marketing with venues and require reusable-bottle access; treat partners as extensions of your brand.
- Connect to local economy and pride. Tie your brand to signature industries (e.g., bourbon) and community identity; showcase how reliable water enables growth.
- Conduct a small group review of your CCR before wide distribution; this helps take in feedback from nontechnical readers and improves the overall report.

Next: Translate brand momentum into management discipline, defining practical key performance indicators (recognition, preference, trust, satisfaction, refill-station usage, partner reach) and cadence (monthly pulse checks, annual brand studies). Customer trust isn't a feeling; it's a measured outcome that can be improved over time.

To prepare this chapter, the author consulted

- Louisville Water presentations and journal materials by Kelley Dearing Smith, including "Louisville Pure Tap," bottle-filling stations, brand journalism approach, research findings, and economic development ties;
- US Environmental Protection Agency's Customer Confident Report Rule found on the agency website at www.epa.gov/ccr/consumer-confidence-report-rule-revisions;
- The International Bottled Water Association and Beverage Marketing Corporation for national bottled-water consumption trends; and
- The Washington State Department of Health Consumer Confidence Report (CCR) site found at https://doh.wa.gov/community-and-environment/drinking-water/regulation-and-compliance/ccr-reports.

Chapter 8

Customer Service as a Profession

What do terms such as *ratepayer* or *call center* say about how we treat the people we serve?

We have noted thus far that customer service is a responsibility of all within the clean water community, from engineering and construction to lab managers and directors. To perform this service well, it must be molded into everyone's job. It is the culture of an organization that emphasizes the community it serves.

As we move into focusing on the most impacted type of role that is front and center in the customer experience, let's dig into how the "customer" is first defined.

The words a utility chooses to describe its customers reflect its values. As tools and technologies evolve, language should, too. Each term the water industry has used over the years from *ratepayer* to *call center* captures a moment in utility culture.

To call customers ratepayers sounds harmless, but it reduces people to bill recipients and puts the utility, rather than the customer, at the center of the relationship. Service becomes transactional: use first, pay later, and talk to someone only if something breaks. That interaction is often viewed as one department's job rather than a shared, organization-wide commitment.

Referring to water customers as homeowners conveys more respect, but it ties access to water service to the name on the deed. Renters, caregivers, and multifamily tenants—often the people directly affected by usage and leaks—can be excluded from key conversations or blocked from receiving timely information.

The variety of terms in use doesn't apply only to utility customers; from *call center agents* to *customer service advisors*, utility employees who support customers are viewed through their words and actions as well as the terms used to describe them. At many utilities, an invisible hierarchical system exists in the treatment of clean water professionals, emphasizing the engineering function over face-to-face customer employees.

A utility's call center historically reflected telephones as the prevalent way customers contacted the utility. The term conveys structure and measurability—average handle time, speed of answer, and other metrics—but it can restrict the mindset to closing help desk tickets rather than building relationships. Even with email, web, app, and chat tools, some digital journeys still lead to phone calls, which signals an opportunity to redesign experiences around resolving issues.

These labels shape how we design systems, train staff, and respond to people. Transforming customer service starts with changing how we talk and think about our customers.

A UTILITY TALE

Andrea, a tenant renting her home, noticed her water bill climbing to a level she could no longer afford. When she called for assistance, the utility could not discuss the account because her name wasn't on the property deed. Meanwhile, she had already alerted her landlord to a possible plumbing issue driving the enormous bill.

Despite being the person using water water every day and most affected by the problem, Andrea was excluded from the service relationship because of a policy linked only to ownership. The result: delayed diagnosis, a bill rendered unaffordable, and preventable frustration on all sides. If a situation like this involves a municipal utility, which often occurs, community residents will be paying for that water leak or delayed maintenance long after the bill has been mailed through escalated rates.

METRICS AND RESOURCES

- About 35% of US households are renters (homeownership rate is about 65% in the second quarter of 2025), so utility policies that only recognize owners can miss a large share of users. In all categories of studies that research basic housing needs for rent, utility bills, food, and medicine, tenants struggle at a higher percentage than homeowners.
- Household leaks are common: The average home can waste around 10,000 gallons of water per year, and 10% of homes misuse 90-plus gallons per day—issues a tenant may notice first.
- Communication matters: Industry research finds satisfaction rises when customers understand safety and infrastructure work, and many digital journeys still end in phone calls, pointing to opportunities to fix dead ends.

Sources for this sidebar:

AWWA and Morning Consult. 2024. *Public Perceptions of Tap Water Survey.* www.awwa.org/wp-content/uploads/Public-Perceptions-of-Tap-Water-Survey-24-Slides.pdf (accessed February 26, 2026).

JD Power. 2024. *U.S. Water Utility Residential Customer Satisfaction Study*. Troy, MI: JD Power.
Urban Institute. 2026. *America's Rental Housing*. www.urban.org/tags/rental-housing (accessed February 27, 2026).
US Census Bureau. *Housing Vacancies and Homeownership*. www.census.gov/housing/hvs (accessed February 27, 2026).
US Environmental Protection Agency. *WaterSense*. United States Environmental Protection Agency, www.epa.gov/watersense (accessed February 27, 2026).

DROPS OF KNOWLEDGE

- Use people-first labels. Align system fields and communications with real users, not just property owners.
- Design for relationships, not service tickets. Measure resolution and follow-through alongside speed metrics. Making sure that customers are entering into financial assistance programs is just as critical a metric as how many bills were mailed last month.
- Open the door for authorized agents. Allow tenants or caregivers to access usage alerts and leak notifications with proper permissions by policy.
- Make digital work end-to-end. Fix common dead ends so customers can resolve issues through self-service rather than having to wait for help at hours (post-6:00 p.m.) that may not align with well-established working patterns.

Next: Hiring with customer service in mind—how to write roles, interview for service mindsets, and bring new employees on board for day-one impact.

Chapter 9

Hiring for Customer Service Regardless of Role

When we hire people who are grounded in customer service—no matter their title—we're hiring the ears, eyes, and understanding of the entire utility. Any employee may be approached to answer a basic question, explain a confusing notice, or calm a worried customer. The following scenarios are common.

- A technician at a meter inspection is asked about new rates.
- A maintenance crew member repairing a sinkhole is questioned about a customer bill.
- A finance analyst is approached at a youth soccer game to explain smart meters.

Customer service doesn't live in one department. It lives in moments like these that span the duties of utility staff. Effective leadership sets the tone for the utility so that any employee can answer simple questions, close the loop, and route issues for follow-up.

According to Mike Hemphill, senior director of leadership development for the Clinton Foundation and former director of the Presidential Leadership Scholars, meaning in a conversation emerges most clearly on the third turn. That is, after two back-and-forths, clarity begins.

Think about a knock-knock joke:

—Knock-knock.
—Who's there?
—Banana.

The meaning is derived after you ask, "*Who's there?*" Then, with the "*Banana*" reply, the framework of the conversation has been established, and it's clear that a joke is being told. Your likely response—"*Banana who?*"— continues the pattern. As a result, the punchline makes sense.

What would this same exchange look like if the answer to "*Banana who?*" were "*I like fruit?*" The meaning would be lost; without a mutual understanding that a joke is being told, the outcome probably wouldn't be successful.

- Real conversations happen in turns back and forth whether by phone, in person, or through an online chat.
- Clarity often lands on the third turn (just like a knock-knock joke that "clicks" after "*Who's there?*").
- Every turn after the third can function like a new "third turn" that adds meaning and momentum.
- Each interaction with customers teaches them what your utility is about: the value of water, how bills are calculated, and whether water is safe to drink.

The feelings that last for customers are shaped by the faces, voices, and digital experiences your utility deploys. Even as technology advances, human-led service remains essential for nuance and trust.

A UTILITY TALE

Las Vegas Valley Water District took a practical path to modernizing customer conversations without losing the human touch. Facing a diverse service area with many language patterns, the project team analyzed recorded interactions and frontline observations to understand how customers signal agreement, confusion, or urgency.

The utility employed three strategies for a positive outcome:

- It listened for meaning beyond the words. Analysts catalogued more than a dozen ways customers say *yes*, from brief acknowledgments to longer confirmations. Coaches then trained staff to pause, paraphrase, and confirm next steps when signals were ambiguous.
- It built assistive technology around real talk. Las Vegas Valley Water District's intelligent virtual assistant (IVA) and agent-assist prompts were tuned with those language patterns in mind so that guidance would reflect how people actually speak in southern Nevada.
- It closed the loop visually and verbally. Follow-up texts and emails repeated the next action in plain language ("*Here's what will happen next . . .*"), reducing repeat contacts.

Early results the utility watched for: fewer transfers on common questions (e.g., payment arrangements, start/stop/move service), shorter handling time for straightforward tasks, and better first-contact resolution on multilingual calls while still routing complex cases to humans.

What you can try: Start a one-page lexicon of local "yes/no/unsure" phrases, embed it in utility staff coaching and your IVA speech, and add a closing script that recaps the customer's goal, your action, and when the customer will hear back from the utility.

THE SIGNAL YOU SEND: JOB DESCRIPTIONS

Some job postings read like time capsules: "This entry-level position involves general IT tasks, including operating mainframe scheduling software. Must be able to lift 25 pounds."

What does a candidate hear?

- Legacy technology warning. "Mainframe scheduling software" suggests a decades-old platform that's harder to secure and staff.
- Recruiting headwind. If you want to hire recent grads and career-switchers, unfamiliar legacy requirements shrink the pool.
- Physical filter. "Must be able to lift 25 pounds" for a desk-dominant role may screen out people who would otherwise be a good fit for the job and sends the wrong message.

Consider the following job description for one utility's customer service department: "This position entails performing customer service functions for the department, including handling inquiries and processing payments. Must be able to perform light physical work and lift and carry up to 25 pounds."

This function-heavy language overemphasizes tasks and underplays the characteristics you actually need. Try reframing: "Work directly with customers compassionately and with integrity to resolve questions and complaints while representing the value of water service for our community."

Finally, define "professional communication" inclusively. What does this mean in a multilingual city? And how are certain sentiments expressed? For example, Las Vegas Valley Water District's review of interactions revealed the many ways that customers may express agreement. The utility applies these insights to customer service coaching and artificial intelligence design. The operational takeaway is to train for meaning rather than jargon (Table 9-1).

Table 9-1 Sample "yes/no/unsure" lexicon for coaching and intelligent virtual assistant tuning

Customer Phrase	Likely Intent	Agent Prompt (Plain Language)
"Yeah/uh-huh/ mm-hmm"	Tentative yes/still listening	"Just to confirm, do you want the payment plan starting Friday?"
"Right, right"	Agreement + impatience	"Got it. I'll keep this brief: your next step is . . ."
"Sure, I guess"	Conditional/unsure	"I can do that, or would you rather ___ ?"
"That's fine"	Acceptance without enthusiasm	"I'll recap: today we ___; next, by Tuesday you'll receive ___."

PEOPLE AREN'T ONE TYPE: PERSONALITY AND ROLE FIT

The Enneagram is one way (among many) to describe how people naturally show up at work. The point isn't to label anyone but instead to widen the path for all different personality features to contribute to good customer care. In a modern utility, different personality styles can shine:

- The Individualist shapes customer-help content and chat prompts; spots tone problems in digital flows.
- The Helper excels in inbound empathy and training new hires.
- The Peacemaker listens for emotion and de-escalates conversations.
- The Investigator digs into meter data and usage patterns and crafts clear explanations.
- The Loyalist nurtures key accounts and community partners.
- The Enthusiast energizes proactive outreach and notifications.
- The Challenger handles tough, policy-bound calls with calm firmness.
- The Achiever fine-tunes digital engagement tools for results.
- The Reformer applies policy consistently (e.g., leak credits) and writes clear, fair decisions.

Bottom line: There isn't one way to "sound" or "act" to excel in customer service. From informal, human notes to highly technical deep dives, different minds strengthen the same mission.

METRICS AND RESOURCES

- The median age of water sector employees is about 48; many utilities anticipate 30–50% of staff will be retirement-eligible in the next 5–10 years.
- The US water sector supports roughly 1.7 million jobs across utilities and supply chains, with many roles not requiring a four-year degree.
- Workforce consistently ranks among the top challenges in AWWA's State of the Water Industry surveys, signaling competition for talent and the need for training processes.
- Clear, purpose-led job postings expand candidate pools and improve retention when roles highlight public service, learning pathways, and growth.

DROPS OF KNOWLEDGE

- Hire for multiple types of customer service duties. Matching your characteristics first, duties second. Emphasize empathy, clarity, and follow-through in postings; retire outdated requirements that shrink your candidate pool.

- Define "professional" inclusively. Train for plain language; catalog how customers say *yes* and *no* across languages and dialects; coach according to meaning rather than jargon.
- Connect roles to strengths. Use personality insights to diversify paths: data explainer, policy communicator, de-escalator, digital content shaper. There's room for different styles.
- Refresh job descriptions annually. Remove duties (e.g., how much weight a person can lift, certain technologies, especially outmoded ones) where they aren't essential; add growth paths (analytics rotations, quality assurance, community engagement).
- Broaden recruiting partnerships. Tap high schools, community colleges, veteran transition programs, and local workforce boards to build a skills-diverse bench.

Next: The value of coaching: turning expectations into daily habits through feedback, practice, and support.

BIBLIOGRAPHY

AWWA. 2024. *State of the Water Industry Report.* www.awwa.org/state-of-the-water-industry/ (accessed February 26, 2026).

Kane, J.W., and A. Tomer. 2018. *Renewing the Water Workforce: Improving Water Infrastructure and Creating a Pipeline to Opportunity.* www.brookings.edu/articles/water-workforce/ (accessed February 27, 2026).

US Environmental Protection Agency. 2025. *America's Water Sector Workforce Initiative.* www.epa.gov/sustainable-water-infrastructure/americas-water-sector-workforce-initiative (accessed February 27, 2026).

Chapter 10

Coaching and Being Coached

Coaching—both giving it and receiving it—is the engine of continuous improvement. While water utility teams don't throw touchdowns or nail three-pointers, they do need the same disciplined approach to practice, feedback, and clear messaging that makes athletic programs thrive.

To show how those habits translate, we spoke with Jimmy Stanton, whose portfolio spans training, coaching, and communications as the senior associate athletic director at the University of Oregon.

INTERVIEW WITH JIMMY STANTON

Q: How do you approach coaching and delivering messages to people with different communication styles or experience levels?

Stanton: Tone is critical when delivering messages, and the impersonality of text or email can let tone be misperceived—often negatively. I prioritize letting department leadership mentor and deliver messages among their own staff, but I always place a high premium on face-to-face communication. Respect for team members should never equate to years of experience. Everyone brings a unique perspective and should have their input valued.

That said, less experienced staff may need more context about why we're taking a particular approach. More seasoned staff may know that inherently. Either way, it's on you to make sure they're hearing the message.

Q: What professional example sticks out for you about unified messaging?

Stanton: When Enron collapsed in 2001, the Houston Astros [a Major League Baseball team] needed to remove the company's name from their ballpark and find a new naming rights partner. Two things stood out.

First, empathy and perspective. The Astros did not emphasize their own hardship and instead focused on the thousands of employees who lost their jobs and savings.

EVIDENCE SNAPSHOT

- Managers matter: Roughly 70% of the variance in team engagement is tied to the manager (Gallup 2017).
- Psychological safety drives performance and learning (Google re:Work 2014).
- Coaching meta-analyses show positive effects on performance, learning, and goal-directed behavior (Theeboom et al. 2014).

Sources for this sidebar:

Gallup. 2017. *State of the American Workplace*. Washington: Gallup Press.

Google re:Work. 2014. *Project Aristotle*. rework.withgoogle.com (project link no longer live).

Theeboom, T., B. Beersma, and A.E.M. van Vianen. 2014. "Does Coaching Work? A Meta-Analysis on the Effects of Coaching on Individual Level Outcomes in an Organizational Context." *The Journal of Positive Psychology*, 9:1:1–18.

Second, one-voice strategy. Only team president Pam Gardner spoke publicly. Her clarity and consistency kept the message focused. No mixed messages, no distractions.

That unity helped the organization maintain credibility during a turbulent time. And the lesson applies internally, too. Own a clear voice, or don't be surprised when results fall short.

Q: How do you get everyone on the same message, especially given that you work with students who are not communication pros, like new hires in a customer service center?

Stanton: Clear, concise communication and a coherent message are essential. When students understand the purpose—why it matters to operations, brand, or the student athlete experience—they grasp why we take the approach we do.

Q: How do you create a culture that welcomes feedback without causing defensiveness?

Stanton: Address the actions or mistakes, not the person. I've seen leaders slip into personal attacks; that creates divisive, inefficient environments. I also welcome insight from all levels, so everyone believes they have a role—because they do.

Q: Can you share a lesson from being coached that shapes your leadership today?

Stanton: A supervisor once told me, "Identifying problems is fine; offering solutions is better." I now empower staff to bring issues and potential solutions. That grows problem-solving capacity, which they'll need when they lead.

Q: How do you balance giving direction with empowering your team to act independently?

Stanton: If you hire the right people and they feel supported—especially when things don't work out—they'll determine the best course rather than always coming to me. I set the big-picture landscape (how an issue affects the department, our values, our mission). Frequent, clear communication ensures they have that context. If they don't, that's on me.

DROPS OF KNOWLEDGE

- Coaching is universal. Whether on the field or in the utility office, people need structured guidance and real-time feedback to master their roles.
- Lead with empathy and perspective. A customer without water is a household in crisis. Focus on solutions, not judgment, just as the Houston Astros focused on laid-off Enron employees before themselves.
- One clear voice prevents confusion. In a crisis or during routine policy rollouts, limit the number of spokespeople and stay disciplined in messaging. Utilizing the communications and public-relations professionals in your organization helps keep the message coherent and unified.
- Grow problem solvers, not alarm sounders. Encourage staff to pair issues with potential fixes; it builds capacity and ownership.
- Policies provide consistency; coaching provides adaptability. Empower employees to apply sound judgment to challenging cases and keep your door open for their ideas.

METRICS AND RESOURCES

- First-contact resolution: percentage of issues fully resolved in the first interaction. Formula: resolved on first touch ÷ total contacts.
- Repeat-contact rate (7/30 days): percentage of contacting again about the same issue within 7/30 or 7 days out of 30.
- Promise kept rate: percentage of callbacks/updates delivered on or before the time promised.
- Complaint reopens: percentage of closed complaints reopened within 14 days.

- Time-to-first update (leak notification or water quality incidents): time from start of incident to first public update (set a published service level agreement).
- Post-interaction customer satisfaction: one- to three-question SMS/interactive voice response/web survey tied to the agent/team.
- Quality assurance calibration score: variance between coach and quality assurance assessors on the same calls (lower is better).
- Coaching coverage: percentage of staff with at least one documented one-on-one meeting per month (and percentage with a ride-along/side-seat/headset listening quarterly).

Bibliography for this sidebar:

AWWA. 2024. *State of the Water Industry Report.* www.awwa.org/state-of-the-water-industry/(accessed February 26, 2026).

Edmundson, A. 1999. "Psychological Safety and Learning Behavior in Work Teams." *Administrative Science Quarterly*, 44:2:350–383.

Gallup. 2017. *State of the American Workplace.* Washington: Gallup Press.

JD Power. 2024. *U.S. Water Utility Residential Customer Satisfaction Study.* Troy, MI: JD Power.

IN CLOSING

Jimmy Stanton's playbook confirms that first-rate customer service is built the same way first-rate sports teams are: with intentional coaching, empathetic messaging, and a culture that values every contributor.

COACH'S CLINIC

High Bill + Leak (at the Front Counter or by Phone): The First 60 Seconds

- Greet and stabilize: "*Thanks for bringing this to my attention. I can see why this would be frustrating.*"
- Set direction but keep the business-speak to basic levels: "*Let's look at how you used water and outline next steps.*"
- Three validating phrases: "You *did the right thing by calling; we are here to answer your questions,*" "*I hear how stressful this is,*" or "*We'll sort through this step by step.*"
- Boundary: "*I can't waive charges without a verified repair, but I can, today, start a leak investigation and note your account.*"

- Resolution path: Review the call discussion points and options/solutions outlines.
- Promise follow-up: *"I'll call you by* [time/date] *with the next update, so you don't have to call back."*

Next: Building a team mentality—aligning operations, billing, field work, and digital communications to deliver one customer experience.

Chapter 11

Team Mentality: We Are One

BUILDING A TEAM

How can you build a customer service culture when your teams don't have the same tools?

Customer service doesn't begin and end at the front desk or call center. As we have mentioned throughout this book, service is shaped by everyone in the organization. The challenge is uniting those varied roles into a true team.

Military organizations understand this well. Commander Josh Portzer, director of global operations for the US Navy, notes, "Being in the military, people often appreciate the idea of being part of something bigger than themselves. Treating people like they are part of a family is often a very effective mindset to foster a team."

A team mentality is built on shared purpose and clarity. But in many utilities, siloed systems and conflicting priorities get in the way. When metering, billing, and service groups aren't aligned, it creates gaps that customers can feel.

One way to realign is to plan backward, imagining your desired outcome and then determining the necessary steps to make that outcome a reality. A "future backward" strategy asks, "What would failure look like?"

When a utility helps a customer, in any role, failure often follows a common pattern:

- The customer doesn't know what's happening or what to expect: "*Why are there people in my yard, and why does my flower garden look different?*"
- Multiple teams respond but don't coordinate: "*They* [the utility] *just sent someone out here yesterday.*"
- The problem is resolved technically, but the customer doesn't know it: "*I called about that yesterday—what do you mean it has been fixed?*"
- The customer is told to call someone else or hears nothing at all.

INFORMATION IS POWER

Any utility can experience a billing system outage. Are you prepared to tell your employees that the system is down and that updates to their current balance, payments being posted, or bill availability are impacted as a result?

How should your field technicians close activities in the automated system since they are not being updated in the billing system? How do you explain to customers that your cashiers cannot post payments because, while your payment portals can process payments, they are not updated in the billing system?

Internal communication tools such as Teams messaging can send a unified message notifying employees of the outage. You can use your public information office to post information on your utility's or the city's website. Customers can be confident that you're aware of the billing system problem and you're working to resolve it. Depending on how long the outage might be, it is a good idea to post periodic updates and keep customers and other stakeholders apprised.

Any number of problems can arise for a water utility. How you communicate about issues and keep employees and customers informed shows that you care about their well-being. Admitting things went wrong can help you gain their trust. Unified messaging and open communication help strengthen that trust.

A UTILITY TALE

When the City of Baltimore overhauled its billing cadence, meter infrastructure, and customer-facing systems, it caused widespread confusion. Customers had questions, and field staff were the ones getting asked. At first, meter technicians had no answers. But that changed.

The city developed laminated cards containing FAQs for technicians to carry in their pockets. This simple tool empowered staff to answer common questions and ease concerns in a consistent way. Staff could provide information if the tone started to escalate while preventing ad hoc conversations that could more easily go wrong. Internally, policy and communication improvements reduced redundant inspections and confusion across teams.

It wasn't a technology that made things work or that was innovative; it was cross-training, collaboration, and clarity. And a simple laminated card.

METRICS AND RESOURCES

- Proactive communication is the single most important step water utilities can take to lift satisfaction; when customers say their utility maintains infrastructure well, overall satisfaction rises by about 229 points (1,000-point scale).

- Less than one-third (almost 30%) of customers recalled receiving any communication from their water utility in 2020; only about 5% recalled four or more communications.
- Ofwat's (in England and Wales) C-MeX weights contact-based service at two-thirds of the overall score (split between operational and billing/general contacts) and noncontact experience at one-third, signaling the value of coordinated cross-team responses.

Bibliography for this sidebar:

AWWA. *Value of Water.* www.awwa.org/communications-and-outreach/value-of-water/ (accessed February 27, 2026).

AWWA and Morning Consult. 2024. *Public Perceptions of Tap Water Survey.* www.awwa.org/wp-content/uploads/Public-Perceptions-of-Tap-Water-Survey-24-Slides.pdf (accessed February 26, 2026).

JD Power. 2020. *U.S. Water Utility Residential Customer Satisfaction Study.* Troy, MI: JD Power.

JD Power. 2024. *U.S. Water Utility Residential Customer Satisfaction Study.* Troy, MI: JD Power.

Ofwat. 2025. *Customer Measure of Experience (C-MeX).* www.ofwat.gov.uk/wp-content/uploads/2025/03/Customer-measure-of-experience-C-MeX-PC-definition-2.pdf (accessed February 27, 2026).

DROPS OF KNOWLEDGE

- Train everyone. All employees need foundational customer service knowledge to operate as one unit.
- Plan backward from failure. Define what poor service looks like and then design backwards to prevent it.
- Equip non–customer service staff. Provide quick-reference materials so any staff member can answer common questions in the field.
- Involve your customers. Educational efforts such as citizen academies can turn residents into advocates and extend your team culture outward.
- Team mentality means thinking beyond departments. It's about building a shared vision, where everyone plays a part in delivering trust, clarity, and connection.

Next: Fitting your message to your customers—adapting tone, channel, and timing for renters, owners, businesses, and vulnerable customers.

Chapter 12

Tailoring Your Message for Different Customer Types

A water utility's duty is to serve everyone: different needs, different expectations, but one standard of service.

Providing the public benefit of serving all customers can be a challenge for water utilities. Most states have requirements spread across their municipal governments to enforce this duty.

Building code: Buildings that will house people or have employees working there require water service, and a certificate of occupancy must be issued.

Real estate: Banks and buyers can request proof of water service to close on a transaction.

Tenancy/landlord regulations: While these rules are often codified, enforcements vary with a location's ability to pass information between agencies (like public works and housing). An example of such wording can be seen in Georgia Code, Title 36:(a).

> No public or private water supplier shall refuse to supply water to any single or multi-family residential property for which water has been furnished through the use of a separate water meter for each residential unit on application of the owner or new tenant of such property because of the indebtedness of a prior owner, prior occupant, or prior lessee to the water supplier for water previously furnished to such property.

These three examples of requirements illustrate three customer types: commercial, homeowner, and renter. In addition, customers like landlords, governmental or intracity entities (hospitals, school districts, city hall), and contractors are all profiles with specific needs, motivations, and methods of engagement. While it is a water utility's duty to serve all customers, developing a treatment plan according to these personas allows customers to have their needs met.

When we consider the kinds of organizations that perform this specialization well, we need only look at our energy cousins: the supplier or retail arm of the deregulated market space. Customer experience is best defined as the perception customers have in every interaction with an organization. For your utility, this concept must be molded to meet the preferences and needs of your service base.

CUSTOMER TYPES

What do these types of customers need, and what defines them? Note that while we are simplifying into fairly basic profiles, each of these profiles has sub-profiles that may be affected by other factors, such as age or primary language.

Homeowners

What defines them—They are residential customers who are part of property tax records at a location. The ultimate responsibility for the water bill is either directly through this person or through a homeowners' association.

Why they are contacting the utility—Usually, they are calling about billing, although they may also call about making changes to the property or have questions about irrigation systems.

How they contact the utility—While the sub-profile of a homeowner may determine how they make contact, these customers may use both self-service and customer service agents.

What they need—They need accurate billing that is easy to understand and a self-service experience that is intuitive to navigate. Homeowner customers appreciate billing that briefly explains the charges but at a level that is clear for individuals who may not be familiar with the water billing structure. Visual and spoken representations of the information should be provided on any digital channel, along with language preferences. Digital channels should be ADA (Americans with Disabilities Act) compliant.

Tenants

What defines them—They are residential customers who lease or rent a property from a homeowner. In this case, the tenant may not be on the actual bill; shared meters may produce a bill that is divided out by the landlord. The customer may be responsible for the bill via a landlord–tenant agreement rather than having direct responsibility.

Why they are contacting the utility—They usually have questions about payments. As tenants are not responsible for the ownership of the property, the information they exchange rarely has to do with the property itself but rather with billing, assistance programs, or plumbing issues that may have led to a higher-than-expected bill.

How they contact the utility—While not an entirely firm rule, tenants generally tend to use calling or interactive voice response rather than digital services. This is largely because of the limitations around payment or assistance options from most digital channels.

What they need—They need information about usage and the tie-in between usage and billing. Tenants need payment arrangements and assistance program options that are readily available online.

Customers need to have a voice about billing. Each utility may have different policies on ownership of service (e.g., whether a tenant is ultimately responsible, whether deposits are taken, what information can be disclosed). In any case, if a person (be it the owner or tenant) shows responsibility for a location and has questions on usage, they need information. Lack of information can strip a person of a voice or the ability to find suitable follow-up about repairs or avoid abuse of the property rental system.

Commercial Organizations

What defines them—These are organizations that are engaged in business rather than residential service. Subcategories can be defined as small to medium businesses, industrial accounts, large governmental, or multisite locations (e.g., schools).

Why they are contacting the utility—They make contact for a mix of reasons. Commercial organizations make up the largest water customer accounts and may have questions that range from not receiving a bill on time because of some exception that prevents it, to management of their locations (large underground leaks), to the forecast of growth in terms of constructing new buildings and expanding beyond their existing water service.

How they contact the utility—This type of customer has many communication channels given their size and complexity. First, they may speak directly to an engineering or construction group to discuss their future plans for growth, available service, building plans, and permitting. Second, they may work directly with a specific account manager, which shortens the resolution process. Finally, they may work with maintenance teams to help pinpoint larger underground leaks on main service lines that create downstream pressure and billing issues.

What they need—Unlike the bulk of their residential colleagues, commercial accounts need specific attention that is trained on their business and can be a central point of contact within the utility. While they may interact with several different departments, escalations are common, so having a designated employee or team to handle that account can simplify determining the root cause of an issue. Thus is the criticality of having key account managers within your customer service department. While billing requests are also common for commercial customers, larger commercial facilities often have detailed knowledge of their usage from internal production meters. Since the revenue that is generated by these accounts often helps to maintain the overall infrastructure of a water utility, having dedicated staff in each area that is knowledgeable about the company's specific operations can keep on-time bill payment and system planning in focus. Ultimately, the water bills of large commercial facilities are the revenue of the utility.

PERSONALIZING THE EXPERIENCE

When determining your utility's personalized treatment plan, you must determine what elements and categories deserve their own profile and ensure that your services are driven by the outcome you want from these customers. Both internal and external factors can assist, but your plan must be measured with outcome and privacy in mind. Said more simply, knowing the location of a property may lead the utility to subcategorize it. There is a fine line between customer profiling for personalized treatment and making assumptions about behavior simply from where a property is located; the latter approach is unacceptable. Common subcategories can be created by incorporating the following information.

- Service line or meter size
- Property owner versus account holder
- North American Industry Classification System (NAICS) code for commercial facilities
- Number of meters present at a location
- Geographic information system (GIS) data
- Medical necessity (if captured in your system)
- Payment source history (Have prior payments been made by assistance agencies?)

Public information sources can help a utility create an experience that benefits customers, no matter who they are.

METRICS AND RESOURCES

- About 35% of US households are renters (homeownership rate was approximately 65.7% in 2025) (US Census Bureau 2026).
- About 22% of US residents age 5+ speak a language other than English at home (US Census Bureau 2023).
- US Department of Justice (DOJ) ADA Title II final rule requires state/local websites and apps to meet Web Content Accessibility Guideline 2.1 AA by 2026–2027 (US DOJ 2024).
- Household leaks: average approximately 10,000 gallons/year; 10% waste 90+ gallons/day (USEPA 2026).

Sources for this sidebar:

US Census Bureau. 2023. "Most Americans Speak Only English at Home or Speak English 'Very Well.'" www.census.gov/newsroom/press-releases/2023/language-at-home-acs-5-year.html (accessed March 2, 2026).

US Census Bureau. 2026. "Quarterly Residential Vacancies and Homeownership, Fourth Quarter 2025." www.census.gov/housing/hvs/files/currenthvspress.pdf (accessed March 2, 2026).

US Department of Justice (US DOJ), Civil Rights Division, ADA. Fact Sheet: New Rule on the Accessibility of Web Content and Mobile Apps Provided by State and Local Governments. www.ada.gov/resources/2024-03-08-web-rule/ (accessed March 2, 2026).

US Environmental Protection Agency (USEPA). 2026. Statistics and Facts. www.epa.gov/watersense/statistics-and-facts (accessed March 2, 2026).

DROPS OF KNOWLEDGE

While no customer is the same, categorize your customers to help determine which functions in your organization can be personalized for those profiles.

- Focus on outcomes. If you want people to enroll in e-Bill, determine the profile of the customer who is likely to enroll. Don't blast customers with useless information because like the boy who cried wolf, when you really need them to respond, they may not.
- Digital experiences are often designed as one size fits all. When determining what information to show on your website, base your decisions on what your customers will do there. If 20% of your customer base and only 2% of those who are past due on a bill log in to their account online, consider the investment you want to make on a top-tier self-service site and invest in the outbound service and proactive work that can be done to assist.
- Commercial customers look for more personalized service because of their size or complexity. If you are trying to lower costs in a contact center, focus on most calls and not the 10 accounts that have 50 meters apiece. For that, focus on key account managers.
- Monitor, monitor, monitor. As times change, so do customers. Profiles are not just a one-time effort; they should be revisited yearly to see if new subcategories should be created.

Next: Designing programs that work within state frameworks and local policy.

Chapter 13

Affordability, Accessibility, and the Rules That Impact Customers

Affordability and accessibility are complex, interconnected topics. Even the most customer-focused utility operates within legal boundaries that vary by state. Legal concepts often shape what is (and isn't) possible: home rule, Dillon's rule, and state-specific funding rules. These frameworks influence how policies on service, rates, penalties, and assistance can be designed (see sidebar).

HOME RULE VERSUS DILLON'S RULE

Home rule: The state delegates broad authority to local governments for matters of local concern.

Dillon's rule: Courts interpret local powers narrowly unless the state legislature clearly grants broader authority. Dillon's rule is a principle of interpretation, not a description of local power itself.

What Drives Affordability?

We consider four aspects that shape outcomes for customers and utilities: historic context, rates, penalties and fees, and affordability programs.

HOW YESTERDAY'S POLICIES SHAPE TODAY'S OUTCOMES

The roots of today's access and affordability gaps reach back to historical housing and lending policies that limited investment in certain neighborhoods for decades. According to the Federal Reserve's archive, alongside the mandate of the Federal Housing Administration to ensure "economically sound" loans arose a fundamentally inequitable situation. As a result, until President Lyndon B. Johnson's Civil Rights Act of 1968 (also known as the

Fair Housing Act), this ambiguous interpretation of what is or is not "sound" impacted not only loans but also ongoing infrastructure issues (Federal Reserve History 2023).

Those patterns left some communities with older housing stock, less resilient infrastructure, and higher exposure to environmental stressors. Research now links these conditions to greater health risks (e.g., respiratory and cardiovascular problems) and higher household costs over time. Why does this history matter to today's decisions? The same areas affected by long-term disinvestment often see higher past-due balances, leak-driven high bills, and greater risk of service interruption. Affordability therefore is not only about ability to pay; it is also about how rate design, fee policy, communication, assistance, and data practices either reduce or unintentionally reinforce those disparities.

BY THE NUMBERS

- 24%: Increase in combined US water and sewer bills since 2019 (Bluefield Research 2025).
- 12–19 million: US households estimated to lack affordable access to water service (USEPA 2024).
- ≥2 million: Number of people in the United States without running water or a working toilet at home (DigDeep and US Water Alliance 2019).
- $8.6 billion per year: Estimated national economic cost of the water access gap (DigDeep and US Water Alliance 2019).
- 80%: Share of US adults with home broadband: Adoption is much lower in the lowest-income brackets (Gelles-Watnick 2024).

Rates: Pressure on Household Budgets

Because bill burden tends to cluster where household incomes and housing quality are lowest, rate design choices directly influence whether those customers fall behind or stay connected.

Nationally, combined water and wastewater bills have risen substantially in recent years. Utilities can soften the effect by pairing any needed rate changes with targeted tools, such as income-based discounts through the use of private or nonprofit funds, leak-adjustment credits that favor permanent fixes and support for plumbing services, or working with legislators to begin the often tedious process of adjusting state-led wording that restricts how discounts can be performed within the municipality.

PENALTIES AND FEES: WHEN UNIFORM RULES HAVE UNEQUAL EFFECTS

Uniform penalties can produce unequal outcomes in areas where past-due accounts are concentrated. During the COVID-19 pandemic emergency (2020–2021), many states and cities temporarily paused shutoffs and, in some cases, reconnected customers (see Portland Water Bureau 2020 for one city's plan). As protections were lifted, utilities resumed standard practices, prompting a necessary question for customer-centric organizations: How can we design collection policies that protect public health, encourage repayment, and avoid compounding hardship?

AWWA policy views discontinuance for nonpayment as a last step after notice, and reasonable alternatives (payment plans, assistance) are offered. That remains a useful foundation, but in households experiencing persistent distress, penalty-first approaches can undermine long-term outcomes (AWWA 2026). (See sidebar for details on state-specific rules.)

STATE-SPECIFIC RULES THE NUMBERS

Some states, such as North Carolina, Alabama, Montana, New Jersey, and Wisconsin, operate with wording that may restrict the ability to enact discounts based on age, income, or other individual-specific criteria:

> "All such charges shall be uniform for the same type, class, and amount of use"—*Code of Alabama § 11-50-121*

> "The rates shall be uniform for like service in all parts of the municipality"—*Wisconsin Statutes § 66.0809*

> "[S]hall. . .be uniform throughout the district for the same type, class and amount of use or service."—*N.J. Stat. § 40:14B-22.*

CONTEXT: THE PAYMENT PLAN MYTH

Extended installment plans alone rarely resolve chronic bill distress. For customers already behind, layering old debt on top of current bills often fails unless paired with real cost relief like leak repair, bill credits, or income-based rates. The lesson from arrearage management efforts in other utility sectors is instructive: Completion rates are modest at best unless programs combine flexible terms with paths to reduce or forgive debt.

AFFORDABILITY PROGRAMS: ASSISTANCE INSTEAD OF PENALTIES

Programs that reduce bills and fix the drivers of high usage tend to perform better for households in persistent distress, and they can stabilize utility revenue. For example, income-based billing with automatic, time-based debt forgiveness is tied to on-time payments as illustrated by the City of Philadelphia's Tiered Assistance Program (or TAP) (Chestnut 2023). The City of Philadelphia saw a 38.2% difference in collection rates with customers who are actively enrolled in TAP versus those who are not.

As another example, holistic support is provided that includes bill discounts, crisis assistance, and free leak repairs for income-qualified homeowners. Some utilities have also seen success when piloting targeted relief for small businesses during the pandemic as the Portland Water Bureau did. These approaches emphasize assistance rather than accumulation of penalties and fees.

DATA STRATEGY: MEASURE THE CUSTOMERS YOU REACH AND WHAT CHANGES

Advanced metering infrastructure provides detailed usage information, but many utilities lack a parallel data strategy for affordability and outcomes. Two hurdles get in the way: identifying and segmenting households so outreach engages those most at risk and measuring whether assistance leads to durable bill stability. Digital access is another barrier as many low-income households do not have reliable home broadband, so web-only enrollment can miss the very customers you want to reach.

Offer multiple application channels, track completion and retention, and publish simple dashboards so leaders and teams can iterate quickly.

EQUITY KEY PERFORMANCE INDICATORS TO TRACK (STARTER LIST)

- Shutoffs per 1,000 accounts by neighborhood (trend over time and location)
- Median arrears and days-past-due data by neighborhood
- Program reach and completion rates by income band/zip code/language
- Leak-repair response time and post-repair bill change

(Notice readability [grade level] and language coverage; inbound call volumes after notices go out.)

A UTILITY TALE

Durham, N.C.: Rethinking Fines, Fees, and Communication

Starting Point and Partners

Durham got involved in a National League of Cities program: Cities Addressing Fines and Fees Equitably, with support from partners that included JPMorgan Chase and the National League of Cities Institute for Youth, Education, and Families.

Under the leadership of former customer and finance manager Heidi Hackett, the team set out to understand who was bearing the bulk of fines and fees related to city services and where those costs were clustering. The goals were practical: Reduce avoidable harm, improve collections, and increase trust, especially in parts of the service area experiencing persistent financial stress.

What the Analysis Found

The team analyzed where past-due balances, late fees, and disconnections were concentrating; reviewed the wording and timing of notices; and listened to feedback from frontline staff and customers. Three issues surfaced repeatedly:

- Clarity: The "last day to pay" and actual shutoff date were not obvious at first glance; total amounts due were easy to misread.
- Assumed literacy and language: Notices used formal phrasing and presumed customers' comfort with bureaucratic terms.
- Missing pathways to help: Letters focused on consequences and left assistance options obscure or entirely absent.

Policy and Fee Changes

Durham piloted targeted policy adjustments to reduce barriers where past-due balances clustered. The city removed a disconnection fee for a customer group to improve the odds that the customer group could direct limited funds toward restoring and maintaining service. This was paired with a clearer on-ramp to help so that the policy change translated into action rather than another confusing step.

The correspondence redesign replaced passive language and dense paragraphs with direct, active phrasing and a single primary action (Figure 13-1). These are some of the specific improvements:

- A bold, plain-language statement of the total amount due and the exact date by which action is required
- A prominent "Need help?" box listing the Water Hardship Fund and other assistance options, with a short website address and phone number
- Simplified, step-by-step instructions for payment and for requesting assistance

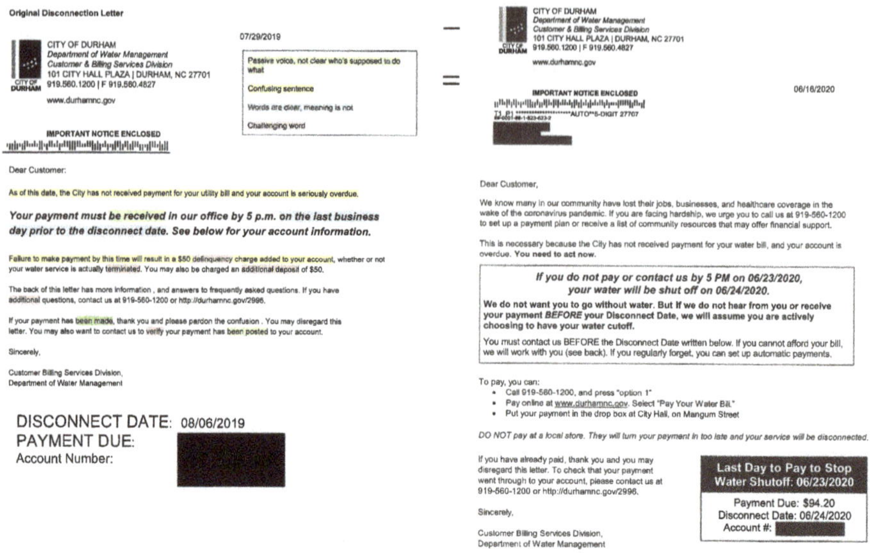

Original Disconnection Letter

07/29/2019

CITY OF DURHAM
Department of Water Management
Customer & Billing Services Division
101 CITY HALL PLAZA | DURHAM, NC 27701
919.560.1200 | F 919.560.4827
www.durhamnc.gov

Passive voice, not clear who's supposed to do what

Confusing sentence

Words are clear, meaning is not

Challenging word

IMPORTANT NOTICE ENCLOSED

Dear Customer:

As of this date, the City has not received payment for your utility bill and your account is seriously overdue.

Your payment must be received in our office by 5 p.m. on the last business day prior to the disconnect date. See below for your account information.

Failure to make payment by this time will result in a $50 delinquency charge added to your account, whether or not your water service is actually terminated. You may also be charged an additional deposit of $50.

The back of this letter has more information , and answers to frequently asked questions. If you have additional questions, contact us at 919-560-1200 or http://durhamnc.gov/2996.

If your payment has been made, thank you and please pardon the confusion . You may disregard this letter. You may also want to contact us to verify your payment has been posted to your account.

Sincerely,

Customer Billing Services Division,
Department of Water Management

DISCONNECT DATE: 08/06/2019
PAYMENT DUE: [redacted]
Account Number:

—

=

CITY OF DURHAM
Department of Water Management
Customer & Billing Services Division
101 CITY HALL PLAZA | DURHAM, NC 27701
919.560.1200 | F 919.560.4827
www.durhamnc.gov

IMPORTANT NOTICE ENCLOSED

T1 P1 **********************AUTO**5-DIGIT 27707

[redacted]

06/16/2020

Dear Customer,

We know many in our community have lost their jobs, businesses, and healthcare coverage in the wake of the coronavirus pandemic. If you are facing hardship, we urge you to call us at 919-560-1200 to set up a payment plan or receive a list of community resources that may offer financial support.

This is necessary because the City has not received payment for your water bill, and your account is overdue. **You need to act now.**

If you do not pay or contact us by 5 PM on 06/23/2020, your water will be shut off on 06/24/2020.

We do not want you to go without water. But If we do not hear from you or receive your payment *BEFORE* your Disconnect Date, we will assume you are actively choosing to have your water cutoff.

You must contact us BEFORE the Disconnect Date written below. If you cannot afford your bill, we will work with you (see back). If you regularly forget, you can set up automatic payments.

To pay, you can:

- Call 919-560-1200, and press "option 1"
- Pay online at www.durhamnc.gov. Select "Pay Your Water Bill."
- Put your payment in the drop box at City Hall, on Mangum Street

DO NOT pay at a local store. They will turn your payment in too late and your service will be disconnected.

If you have already paid, thank you and you may disregard this letter. To check that your payment went through to your account, please contact us at 919-560-1200 or http://durhamnc.gov/2996.

Sincerely,

Customer Billing Services Division,
Department of Water Management

Last Day to Pay to Stop Water Shutoff: 06/23/2020

Payment Due: $94.20
Disconnect Date: 06/24/2020
Account #: [redacted]

Source: City of Durham, N.C.

Figure 13-1 Communication rewrite (before → after)

- Translation and readability adjustments to reduce the literacy burden
- Consistent placement of the key dates so customers and staff can reference the information quickly

Assistance Paths and Funding Alignment

Because Durham's water hardship funds originate in the general fund, the program could be administered via the water fund to streamline access. Customer service staff were trained to connect callers facing shutoff risk directly to hardship screening. The city also moved the water hardship and payment arrangement forms online—while maintaining phone and in-person options—to widen access and reduce friction. In parallel, the team explored additional state and federal programs and public–private partnerships to expand the menu of help available to customers.

Governance, Training, and Workflow

Durham's approach emphasized cross-departmental alignment among customer service, finance, legal, and communications so that policy, letters, and call-center scripts matched. Staff received refresher training on empathy, plain-language communication, and consistent documentation. Workflows were updated so that customers who called about disconnection notices encountered an "assistance-first" path before any penalty discussion.

Measuring Results

To know whether the changes were working, Durham began tracking a simple set of indicators: disconnections per 1,000 accounts by neighborhood, median arrears and days past due, assistance program take-up and completion, and call volumes after notices. Early internal feedback indicated fewer callbacks from customers confused about dates and amounts due, greater awareness of hardship options, and improved follow-through among customers who entered an assistance path.

METRICS AND RESOURCES

Durham: Transferable Steps You Can Adopt

- Start with a map: Visualize where arrears, fees, and disconnections cluster. Have some facilitated conversations to develop an approach strategy; this same information will need to be used with regulatory agencies, including city councils, when making any changes.
- Rewrite the letter: one action, one amount due, one clear date, plus a prominent help path.
- Align policy with intent: Remove or reduce fees that block reconnection for targeted groups by preparing your customer information and making a data-based argument for change.
- Broaden access: Put assistance and payment-arrangement forms online and keep nondigital paths.
- Measure for learning: Track shutoffs/arrears, assistance enrollment/completion, and post-repair bill changes.

FROM CONCEPT TO CULTURE

Rules set boundaries, but service culture determines how we operate inside them. Durham's experience shows how a service-first perspective emphasizing clarity, access to help, and targeted policy changes can interrupt cycles of distress while protecting public health and stabilizing revenue.

KEY TAKEAWAYS

- Build assistance before penalties. Pair payment plans with bill reductions (income-based rates), leak repairs, and time-based debt forgiveness.
- Fix the letters, fix the experience. Make dates, amounts, and assistance options unmistakable and verify readability.
- Mind the digital divide. Offer multiple enrollment channels and simplify forms to reach households that don't have reliable broadband.

- Measure what matters. Track whom you reach, who completes assistance, and who stays current 6–12 months later; publish simple dashboards, and be ready to pivot to new approaches if outcomes don't improve.

Next: Nudging adoption of helpful customer service programs, from payment options to leak alerts.

WORKS CITED

AWWA. 2026. *Discontinuance of Water Service for Nonpayment.* www.awwa.org/policy-statement/discontinuance-of-water-service-for-nonpayment/ (accessed February 27, 2026).

Bluefield Research. 2025. *Municipal Utility Water Rates Index 2024: Drinking Water&Sewer.* www.bluefieldresearch.com/research/u-s-municipal-drinking-water-sewer-annual-utility-rate-index/ (accessed February 27, 2026).

Chestnut, M.R. 2023. *Hearing Officer Report: In the Matter of the Philadelphia Water Department's Proposed Changes in Water, Wastewater and Stormwater Rates and Related Charges, for Fiscal Years 2024–2025.* www.phila.gov/media/20230530162301/2023-Hearing-Report-May-30-FINAL-with-Appendix.pdf (accessed February 27, 2026).

DigDeep and US Water Alliance. 2019. *Closing the Water Access Gap in the United States: A National Action Plan.* www.digdeep.org/s/Dig-Deep_Closing-the-Water-Access-Gap-in-the-United-States_DIGITAL_compressed-2hyx.pdf (accessed February 27, 2026).

Federal Reserve History. 2023. *Redlining.* www.federalreservehistory.org/essays/redlining (accessed February 27, 2026).

Gelles-Watnick, R. 2024. "Americans' Use of Mobile Technology and Home Broadband." www.pewresearch.org/internet/2024/01/31/americans-use-of-mobile-technology-and-home-broadband/ (accessed February 27, 2026).

Portland Water Bureau. 2020. "Media Advisory: City of Portland Utilities Pledge $1 Million for Small Businesses Impacted by COVID-19." www.portland.gov/water/news/2020/6/3/media-advisory-city-portland-utilities-pledge-1-million-small-businesses (accessed February 27, 2026).

US Environmental Protection Agency. 2024. *Water Affordability Needs Assessment: Report to Congress.* www.epa.gov/system/files/documents/2024-12/water-affordability-needs-assessment.pdf (accessed February 27, 2026).

Chapter 14

The Marketing and "Sales" Side of Customer Service

Why terms like *marketing* and *sales* belong in the water industry...

Personalization and a utility-wide view of service require a living approach. Years ago, terms like *marketing*, *sales*, and *retention* felt out of place in the water industry. But the goals are the same as in competitive markets: Help customers make good choices, reduce avoidable costs and surprises, and build trust so people stay engaged. JD Power's *Water Utility Residential Customer Satisfaction Study* measures satisfaction across six factors—quality and reliability, price, conservation, billing and payment, communications, and customer service—exactly where day-to-day marketing and sales behaviors show up (JD Power 2023).

Create a Sales Culture (Without It Actually Being "Sales")

Start at hiring. If you need enrollment growth in programs (autopay, paperless/eBill, leak alerts), say so in the posting. Look for a persuasive, customer-teaching mindset across all channels (phone, walk-in, chat, web form, in the field). In interviews, test candidates on explaining options clearly and asking for the next step ("While I'm looking at your account, would you like to have leak alerts turned on?"). Train new hires to ask gentle, value-forward questions, then reinforce them in quality assessment. Recognize solid soft sells, even when the customer declines. Treat these behaviors as routine, not a one-off campaign.

Acquire, Develop, Retain: What That Means for Regulated Utilities

In the water industry, the word *acquisition* is smooth move-ins and clean move-outs. *Development* is growing participation in programs that lower risk (autopay, leak alerts) and friction (paperless). *Retention* is preventing dissatisfaction from turning into complaints, late payments, or political pressure. Adoption plateaus without ongoing prompts, so weave small, timely requests into everyday journeys and not just into launch campaigns.

Provide Choices and Earn Attention

Energy providers talk about the customer value experience, matching offers to habits and preferences. For the clean water industry, that means pairing local priorities (water budgets, reuse, conservation, leak prevention, builder outreach) with multichannel options—landline, web, cellphone, text—and timed prompts.

Two practical reminders:

- Trust beats tech. Behavioral "nudges" work when they are transparent and helpful. In a recent paper by Katrina Jessoe and colleagues (Jessoe et al. 2020), "home water reports" from households that received water-use feedback (social norm + tips) reduced their electricity use by approximately 1.5–2% in summer, especially at times when water-using appliances run. The authors interpret this as a water-to-energy awareness channel: Learning that water use carries energy costs prompts complementary conservation that shows up on the electric bill.
- Make the helpful thing the easy thing. Don't bury autopay or leak-alert sign-up. Put the offer where the need appears: e.g., after a high bill, during move-in, following a leak ticket.

Identify Dissatisfaction Early (and Act on It)

Your contact data will show where there is friction: patchy communications during planned/unplanned water shut-offs, unclear bill design, surprises from meter changeouts, and hard-to-use websites. Start with high-impact, low-effort fixes and stand-up cross-functional "issue squads" for outages and restorations (customer service, field, dispatch, communications, legal, IT). Then close the loop publicly when you improve something.

Amplify Your Advocates

Nothing sells your programs like customers. Nielsen's global research shows 88% of people trust recommendations from people they know, the most-trusted channel by far (Nielsen 2021). Make it easy for happy customers to speak up: Ask for a short quote, initiate a survey after a successful leak-resolution or high-bill coaching call, capture 2-minute "how this helped me" videos for your site and council briefings, or recruit education program graduates as community ambassadors.

Know When to Stop (or Change Course)

Great marketers quit using tactics that don't work. In utilities, that often means retiring bill stuffers that few customers act on and shifting those dollars to targeted outbound for small, high-impact segments (e.g., repeated late payments). One Paymentus client reported approximately $343,000 per year in postage and print savings after 55,000 accounts went paperless, funds you can reallocate to proactive outreach (Paymentus 2022).

Map the Work, Measure What Matters

Don't wait for a customer-information service project to map and fix customer issues: move-in/move-out; exception handling (high bills, unreadable meters, stuck field orders); and field timelines (reconnects, meter change-outs, leak checks). Set up metrics you'll actually use.

- Program adoption: eBill enrollment, autopay/automated clearing house (ACH) rate, leak-alert opt-ins
- Experience and communication: percentage of customers notified before planned water shut-offs, first-contact resolution on high-bill calls, and clarity of billing and payment and communications (key factors tracked by JD Power) (JD Power 2023)
- Financial and operations: write-offs as a percentage of revenue
- Exception volume and cycle time; cost-to-serve per contact (live versus online)

Sell the Good Works (Ethically)

When you promote conservation, leak prevention, or budget billing, you are reducing risk and expense for both household and system. Social-norm water programs show approximately 4–5% savings; at scale, that is capacity you don't have to build (Brent et al. 2015, Bernedo et al. 2014, Ferraro and Price 2013). And because water and energy are linked, the US Environmental Protection Agency estimates drinking water and wastewater systems account for about 2% of US energy use and often 30–40% of a municipality's total energy consumption (USEPA 2025).

What This Means for Customer Service Teams

A monthly bill to a "captive" customer by a monopoly is not enough; customers judge you against banking, retail, and telecommunications. Your team's daily conversations are your marketing: how you explain options, when you make the ask, and how you follow up.

PRACTICAL MOVES THIS QUARTER

- Tune your scripts: Embed one relevant offer in each common call path (move-in → paperless + autopay; high bill → leak alerts + usage tips; payment trouble → assistance + plan).
- Fix water shut-off communications: Assign a single owner, publish service level agreements, and test messages on real devices.
- Add a gentle nudge after performing proactive notification measure usage on a randomized sample.

METRICS AND RESOURCES

Metrics to Watch: Checklist

Program Adoption

- eBill (paperless) enrollment rate
- Autopay/ACH enrollment rate
- Leak-alert opt-in rate, time to resolution after alert
- Experience and communication
- Percentage of affected customers notified before planned water shut-offs
- First-contact resolution for high-bill calls
- Customer satisfaction on billing and payment and communications categories per JD Power
- Financial and operations
- Write-offs as percentage of revenue
- Exception volume and cycle time (meter reads, field orders, billing holds)
- Cost-to-serve per contact (live versus digital)

DROPS OF KNOWLEDGE

- Everyone sells value. You may never use the word *sell*, but every role should offer timely, value-based offers that help customers and protect the system.
- Target, test, adjust. Start with high-impact/low-effort fixes, measure, and stop doing what does not work.
- Trust + timing win. Clear information, well-timed communications in the customer's channel of choice, and easy next steps beat big random mail stuffers.
- Showcase your advocates. Word of mouth is the most trusted channel (88% trust people they know) (Nielsen 2021).

Next: How key performance measures help utilities define what good performance looks like with metrics based on what customers actually experience.

WORKS CITED

Bernedo, M., P.J. Ferraro, and M.K. Price. 2014. "The Persistent Impacts of Norm-Based Messaging and Their Implications for Water Conservation." *Journal of Consumer Policy*, 37:3:437–452. https://doi.org/10.1007/s10603-014-9266-0.

Brent, D.A., J.H. Cook, and S. Olsen. 2015. "Social Comparisons, Household Water Use, and Participation in Utility Conservation Programs: Evidence from Three Randomized Trials." *Journal of the Association of Environmental and Resource Economists*, 2:4:597–627. https://doi.org/10.1086/683427.

Ferraro, P.J., and M.K. Price. 2013. "Using Nonpecuniary Strategies to Influence Behavior: Evidence from a Large-Scale Field Experiment." *The Review of Economics and Statistics*, 95:1:64–73. https://doi.org/10.1162/REST_a_00344.

JD Power. 2023. "Water Utilities Counteract Negative Effects of Inflation with Strong Customer Communication Strategies, JD Power Finds." www.jdpower.com/business/press-releases/2023-us-water-utility-residential-customer-satisfaction-study (accessed February 27, 2026).

Jessoe, K., G.E. Lade, F. Loge, and E. Spang. 2021. "Spillovers from Behavioral Interventions: Experimental Evidence from Water and Energy Use." *Journal of the Association of Environmental and Resource Economists*, 8:2: 315–346. https://doi.org/10.1086/711025.

Nielsen. 2021. "Beyond Martech: Building Trust with Consumers and Engaging Where Sentiment Is High." www.nielsen.com/insights/2021/beyond-martech-building-trust-with-consumers-and-engaging-where-sentiment-is-high/ (accessed February 27, 2026).

Paymentus. 2022. *Reducing the Cost to Serve Customers and Days Outstanding Through Digital Transformation*. https://cdn.featuredcustomers.com/CustomerCaseStudy.document/Paymentus_Case_Study_Mortgage_2022.pdf (accessed February 27, 2026).

US Environmental Protection Agency. 2025. *Energy Efficiency for Water Utilities*. www.epa.gov/sustainable-water-infrastructure/energy-efficiency-water-utilities (accessed February 27, 2026).

Chapter 15

Key Performance Measures

Key performance indicators (KPIs) do more than steer day-to-day operations; they help utilities tell a clear story to councils, directors, human resource departments, employees, and the community. We have all heard the cliché, if you can't measure it, you can't manage it. The most effective KPI programs align measures across departments so that the entire organization pulls in the same direction. When KPIs are misaligned, one team might appear to be succeeding in its measurements even though customers are experiencing issues with their water service or billing.

This chapter provides guidance on performance metrics that employees may use to identify their "health" within your utility. Not all may apply to you, but consider how the lack of that measurement may affect how you and others in your utility approach customers.

NOTABLE STATISTICS

- JD Power surveyed 33,476 residential customers across 91 large utilities in its 2025 study; overall satisfaction was 515/1,000 (JD Power 2025).
- Many contact centers still use an 80/20 service level target (80% of calls answered within 20 seconds) and aim for approximately ≤5% call abandonment; tailor this metric to your customers and staffing reality (SAMHSA 2025, Bave-Kerwin 2011).
- An average of 50% of inbound calls to a water utility call center are about billing-related issues, while the numbers can be much higher, given that calls may be identified as "payment" or "questions," which may also be related to billing but not tagged as such (FGUA 2025).

CORE KPI FAMILIES AND STARTER SET

Assets and Metering

- Meter accuracy (revenue meters): AWWA standards typically specify 98.5–101.5% accuracy at normal flows; use a meter-test pass-rate and the share of meters past test life as leading indicators.
- Asset condition rating: Track condition scores and the percentage of assets below your threshold.
- Planned versus reactive maintenance ratio. In a reliability-centered maintenance comparison, reactive maintenance is presented as an approximately 200% cost increase over prevention.
- Main breaks per 100 miles per year and leakage index: Recent benchmarking shows averages near 11–14 breaks per 100 miles depending on pipe cohort and study year.

System Reliability and Service Continuity

Borrow the logic of reliability indices used by electric utilities: System Average Interruption Duration Index/Customer Average Interruption Duration Index (SAIDI/CAIDI)—i.e., system/customer average interruption duration indices. Define your own service interruption KPIs for water (Table 15-1). Track average interruption duration, customer interruption frequency, number of customers affected per incident, minutes lost per customer, and time to restore. Most important is the ability to apply these metrics to specific areas of your service territory because the customers in the same area are often experiencing these interruptions at the same time, and this is a fact that can get lost when reporting is totalized per year. Use International Water Association and AWWA benchmarking definitions for consistent terms (see AWWA 2025).

Customer Operations, Billing, Payments

- Bills issued on time (%)
- Billing accuracy (% error-free bills)
- Estimated-bill rate (%)
- Average cost per bill by segment (residential, commercial)
- Average number of days from meter read to receipt of bill
- Payment channel mix and cost-to-serve (lockbox, interactive voice response system, website, walk-in)
- Paperless/eBill enrollment; autopay/ACH (i.e., electronic transfer of payment) enrollment
- Collections health: % outstanding revenue, % with >2 past-due bills, % active and % broken arrangements, bad-debt %
- Service affordability burden: typical monthly water + sewer bill as a percentage of estimated household income (report median and low-income quintile)

Table 15-1 Service interruption key performance indicators

Key Performance Indicator	Definition	Formula Example
Actual-read billing rate	Share of bills produced with an actual meter read in the cycle	Actual-read bills ÷ total bills × 100
Estimated-bill rate	Share of bills produced using an estimate in the cycle	Estimated bills ÷ total bills × 100
No bills	Active accounts with no produced bill	Total number of active accounts in a given cycle ÷ total bills produced
Bills issued on time	Bills generated and delivered by the scheduled bill date	On-time bills ÷ total bills × 100
Billing accuracy	Bills issued without billing errors (reversals/adjustments)	(Total bills - error bills) ÷ Total bills × 100
Billing-critical field service level agreement	Timeliness for field orders critical to billing	Orders completed within X days ÷ total billing-critical orders × 100
First-contact resolution	Issues resolved in a single contact with no follow-up	Issues resolved on first contact ÷ total issues × 100
Average speed of answer	Average time (seconds) to answer inbound calls	Total wait time for answered calls ÷ answered calls
Abandon rate	Share of inbound calls that disconnect before answer	Abandoned calls ÷ total inbound calls × 100
Service level (e.g., 80/20)	Share of calls answered within a threshold	Calls answered ≤ threshold ÷ total answered × 100
Cost per bill	Average fully loaded cost to produce and deliver a bill	Total billing costs ÷ total bills
Cost per payment (by channel)	Average fully loaded cost to process a payment	Channel payment ops cost ÷ payments (channel)
Paperless (eBill) enrollment	Accounts enrolled in paperless billing	Active eBill accounts ÷ active accounts × 100
Autopay/ACH enrollment	Accounts enrolled in automatic payments	Active autopay accounts ÷ active accounts × 100
>%2 past-due accounts	Accounts with more than two past-due bills	Accounts with >2 past-due ÷ active accounts × 100
Arrangement break rate	Share of payment arrangements that are broken	Broken arrangements ÷ active arrangements × 100
Bad-debt percentage	Write-offs as a share of billed revenue	Write-offs ÷ billed revenue × 100
Nonrevenue water: apparent losses	Consumption not billed due to meter error and theft	Apparent loss volume ÷ system input × 100
Nonrevenue water: real losses	Physical water losses from leaks/breaks	Real loss volume ÷ system input × 100
Main breaks per 100 miles	Break frequency normalized by network size	Annual main breaks ÷ (miles of main ÷ 100)
Planned water-off notification coverage	Affected customers notified before planned outage	Customers notified ÷ customers affected × 100

(continued)

Table 15-1 Service interruption key performance indicators (*Continued*)

Key Performance Indicator	Definition	Formula Example
Percentage of on-time meter reads	What percentage of the service territory has an actual read each cycle (whether via advanced metering infrastructure, automated meter reading, or manual reading)	Actual nonestimated read per cycle ÷ active services per cycle × 100

- Percentage of payment plans successfully completed
- Medical necessity indication as percentage of total active accounts
- Past-due customers: Determine what percentage is actively on a payment or income-qualified program with your utility (or municipality).

Contact Center and Digital Experience

- Average speed of answer and abandon rate: Set targets based on customer expectations and staffing realities (80/20 and approximately ≤5% abandon are common references).
- First-contact resolution
- Total engagements and containment (issues resolved via self-service)
- Customer satisfaction by journey (move-in/move-out, high bill, payment assistance, water shut-off communications).

Water Loss and Energy Intensity

- Nonrevenue water as calculated using the AWWA water-audit methodology (software available for free on awwa.org) splits apparent losses from real losses and produces a data-validity score.
- Real losses (gallons per connection per day)
- Energy per million gallons treated/pumped; track alongside process changes and seasonal demand
- Total energy cost
- Energy demand cost
- Energy cost as a percentage of collected revenue

A UTILITY TALE

This is the story of metric mismatching, a common occurrence. The lesson? It is important to remember not to assign metrics to groups that have no control over them. What follows is a story of how common this practice is.

Joe works in billing. One of his KPIs is the percentage of estimated bills. When reads are missing—perhaps a meter stopped communicating or access was blocked—he can issue a short-term estimate, but an actual read is required within a particular period of time. Joe dispatches a field order to get those reads.

Letizia is in charge of field service. Her KPIs include the percentage of work orders completed and the timeliness of repairs and appointments. Her calendar is full of water turn-ons, reconnections, customer appointments in new developments, and meter replacements for an advanced metering infrastructure pilot test, mostly in a different territory from Joe's pending read-requests.

Result: Joe's estimated-bill rate rises in pockets of the city, while Letizia still meets her productivity goals. Both teams are performing to their own KPIs, yet the customer sees estimated bills and uneven service. The customer now contacts the call center as a result of confusion over the bill, impacting the metrics on customer engagement, number of inbound calls, and handle time. This is a classic sign that metrics are not aligned.

METRICS AND DATA: ALIGNING "WINS"

- Actual-read billing rate (billing + field)
- Estimated-bill rate (shared; target decline)
- Billing-critical field service level agreement; % within X days)
- Customer impact (high-bill callbacks/1,000 bills)

DROPS OF KNOWLEDGE

- Align KPIs across teams so that no one can "win" while the customer loses.
- Use standard definitions so that results are comparable over time and across peer groups.
- Track a few shared outcomes that customers express and that have downstream impacts on on-time, accurate, actual-read bills; clear water shut-off communications; and fast, first-contact resolution.
- Pair KPIs with small process fixes and measure the before and after; stop doing what doesn't work, not after 1 year or one administration, but reassessing constantly, cycle by cycle and quarter by quarter.

Next: A step-by-step approach to developing a comprehensive strategy that keeps stakeholders informed and engaged.

WORKS CITED AND BIBLIOGRAPHY

Alegre, H., J. Melo Baptista, E. Cabrera, Jr., F. Cubillo, P. Duarte, W. Hirner, W. Merkel, and R. Parena. 2016 (3rd ed.). *Performance Indicators for Water Supply Services*. London: IWA Publishing.

AWWA. 2024 (5th ed.). M36, *Water Audits and Loss Control Programs*. Denver: AWWA.

AWWA. 2025. *AWWA Utility Benchmarking: Performance Management for Water and Wastewater.* Denver: AWWA.

Barfuss, S.L. 2023. *Water Main Break Rates in the USA and Canada: A Comprehensive Study.* Logan, UT: Utah State University.

Bave-Kerwin J. 2011. *Contact Center Metrics: Why Metrics Matter and Why.* Colorado Springs, CO: ICMI. www.icmi.com/files/StudentResourcePage/CCMetrics/Slides_VC654.pdf (accessed February 27, 2026).

COPC, Inc. 2024. *Service Level—Are You Measuring It the Wrong Way?* www.copc.com/service-level-are-you-measuring-it-the-wrong-way/ (accessed February 27, 2026).

Fiserv. 2024. *Expectations & Experiences: Digital Billing and Payments—Fast Facts.* www.fiserv.com/content/dam/fiserv-ent/final-files/marketing-collateral/documents-handouts-flyers/expectations-and-experiences-digital-billing-and-payments-fast-facts-0724.pdf (accessed February 27, 2026).

Florida Governmental Utility Authority. 2025. January 16, 2025, Board Meeting Packet Supplemental. www.fgua.com/docs/board/packets/January%2016,%202025%20FGUA%20Board%20Meeting%20Package%20Supplemental.pdf (accessed February 27, 2026).

JD Power. 2025. *U.S. Water Utility Residential Customer Satisfaction Study.* Troy, MI: JD Power.

Pacific Northwest National Laboratory. *O&M Best Practice Issue Discussion: Maintenance Approaches.* www.pnnl.gov/projects/om-best-practices/maintenance-approaches (accessed February 27, 2026).

Substance Abuse and Mental Health Services Administration (SAMHSA) 2025. *988 Lifeline Performance Metrics.* www.samhsa.gov/mental-health/988/performance-metrics (accessed February 27, 2026).

WSSC Water. 2025. “Meter Testing.” www.wsscwater.com/customer-service/rates-and-billing/know-your-household-water-usage/meter-testing (accessed February 27, 2026).

Chapter 16

Communications

Conveying a message of great service is only half the job; people have to hear it, see it, and believe it. This chapter turns communications into a repeatable utility practice and describes how one utility made it both memorable and measurable.

Water and wastewater teams can't be silent operators anymore. Trust must be earned in public early, often, and with facts that matter to customers. Use communications as a disciplined management function through planning, resourcing, and measuring, just like you would with any other program.

Throughout this book, we have discussed exceptional cases of customer service. Any utility, regardless of size, can benefit from a sound communications strategy. Communication extends beyond media, letters, or online social forums; it is a strategy that your utility employs when trying to make sure your community is informed.

GETTING STARTED

- Step 1: Draft a one-page plan, an initial communications playbook; identify three priority audiences and three SMART (specific, measurable, achievable, relevant, time-based) objectives.
- Step 2: Build a message map and a 6-week editorial calendar; assign owners.
- Step 3: Publish regularly on owned channels; apprise local media.
- Step 4: Host one community touchpoint (e.g., tour, pop-up table, webinar).
- Step 5: Review metrics, stop using what underperforms, and focus on what works.

Table 16-1 Stakeholder audience matrix

Audience	Core Promise	Proof Points	Action (Specific)
Homeowner	Your water is reliable, safe, and fairly priced.	Annual testing; 24/7 monitoring	Annual water quality report, comparison studies
Renter	We help you avoid surprises.	Usage notifications; flexible billing options	Add a text alert number.
Elected official	We spend wisely and report results.	Capital plan; performance dashboard	Attend quarterly briefing.

Most customer service crises begin as communication failures.

- Surprise rate increases
- Unclear watering rules
- Technical notices that read like regulatory filings
- Service interruptions announced after the fact

When customers are confused, they call.
When they are surprised, they escalate.
When they feel uninformed, they distrust.

According to the 2023 JD Power *U.S. Utility Residential Customer Satisfaction Study*, communication clarity and ease of understanding are consistently among the strongest drivers of customer satisfaction in utilities.

At the same time, research from the Edelman Trust Barometer (2024) shows that local institutions, including utilities, retain higher trust than national entities, but that trust declines quickly when transparency is perceived as lacking.

For water utilities, communications is not marketing. It's a form of service delivery (Table 16-1). The purpose of communications in customer service is simple:

- Prevent unnecessary contacts
- Reduce misinformation
- Improve compliance
- Increase understanding
- Build predictable trust

When communications work, customers are not surprised.

COMMUNICATIONS AS A CUSTOMER SERVICE TOOL

This chapter is not about building a communications department. It is about equipping customer service leaders to use communications strategically.

Three principles anchor this work:

Prevent before you respond: Customer service teams often measure success by response time and resolution rate. Those matter, but the most powerful metric is prevention.

Ask: Do customers understand rate changes before they appear on the bill? Do they know watering rules before enforcement begins? Do they understand what causes service interruptions?

A study from AWWA found that utilities that proactively communicate during rate adjustments report fewer complaint-driven contacts compared to utilities that rely solely on bill inserts (AWWA 2022).

Practical actions are listed here.

- Create one-page plain-language explainers for complex issues.
- Announce seasonal rules weeks before enforcement.
- Use short video or visual formats for high-confusion topics.
- Publish FAQs before confusion spikes.
- If customers can repeat your message accurately, your communications are working.

Tone Is a Service Decision

Utilities operate in regulated, technical environments. That can unintentionally create rigid messaging.

But tone influences behavior. Two utilities (The Town of Flower Mound, Tex., and Southern Nevada Water Authority) created YouTube campaigns with humor to explain water rules and conservation practices and drove engagement. Rather than issuing formal directives, they

- simplified one clear rule,
- used familiar cultural references,
- presented staff as relatable, and
- delivered the message before peak season.

The result? Higher engagement, broader reach, and increased community awareness were noted. The lesson is not "be funny." The lesson is that technical rules are easier to follow when they are human.

Research from the Pew Research Center (2025) shows that audiences are significantly more likely to engage with content that is relatable and emotionally resonant rather than strictly informational.

For customer service teams, this translates to

- avoiding punitive framing,
- replacing regulatory jargon with everyday language,
- matching tone to community culture, and
- reviewing scripts for clarity, not just compliance.

Humor lowers defensiveness. Clarity drives compliance.

Risk Register and Crisis Ladder

These are the steps in addressing communication: list likely scenarios (main break, boil-water notice, service line hit, rate confusion); pre-draft first statements, FAQs, and spokesperson notes; define who activates, who approves, and what the first three actions are (Table 16-2).

Table 16-2 Understanding the roles and responsible, accountable, consulted, and informed chart of communication

Task	Responsible	Accountable	Consulted	Informed
Editorial calendar	Communications manager	Director	Operations, customer experience	Executive team
Message approvals	Communications manager	Director	Legal	General manager/ Board
Crisis statements	Public Information Officer	General manager	Legal, Operations	Board
Metrics report	Analyst	Director	Customer experience, Operations	Executive team

Table 16-3 Metrics and monitoring

Leak alert enrollments	Outcome (action)	+20%
First-contact resolution	Outcome (quality)	≥85%
News post reads	Output (reach)	+30%
Negative media ratio	Outcome (reputation)	<10% of clips

Community Touchpoints

Schedule standing, predictable opportunities for dialogue, such as during operational tours, library hours, post religious gatherings, school talks, neighborhood events, and pop-up tables at markets. Show up where people already gather.

Measurement and Feedback

Track both outcomes (trust, satisfaction, contact rate, payment behavior) and outputs (reach, opens). Hold a 30-minute monthly review to stop doing what doesn't work and scale up what works (Table 16-3).

A UTILITY TALE

Denver Water: Newsroom Thinking, Community Voice

Denver Water organized its communications like a newsroom, with writers, editors, and video, publishing on the *TAP* news site and social channels. Beyond press releases, the utility built a steady cadence of useful human-interest stories: lead service line replacements, water-wise landscaping, and staff profiles.

Their 2024 parody, "I Water That Way," sung by the Splashstreet Boys (Figure 16-1), turned summer watering rules into a catchy Backstreet Boys homage shot on an iPhone with 1990s outfits. The video landed on NBC's *Today* show, approached 200,000 views, won the Bronze Anvil award from

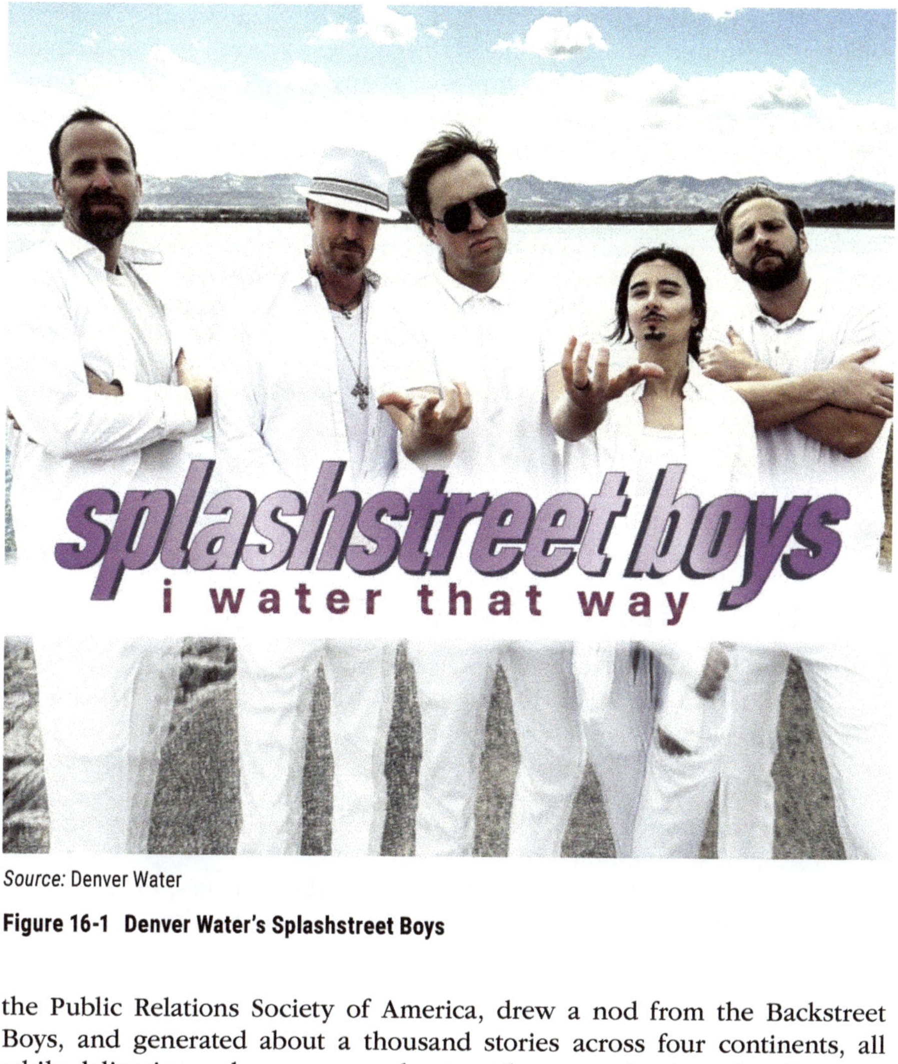

Source: Denver Water

Figure 16-1 Denver Water's Splashstreet Boys

the Public Relations Society of America, drew a nod from the Backstreet Boys, and generated about a thousand stories across four continents, all while delivering a clear message about outdoor watering in a dry climate.

MAKE THE DENVER WATER STORY WORK FOR YOU

1. Define one simple rule to promote (e.g., watering days).
2. Write a 30-second script with a single "remember this" line.
3. Produce with what you have (e.g., phone + captions).
4. Publish natively on your strongest social channels.
5. Link back to a short, plain-language explainer page.
6. Measure: 3 reach metrics + 3 action metrics; report in 2 weeks.

Why it worked: Leadership backed calculated creativity, the core message was clear, and the team went to the channels where residents actually spend time. You can adapt this approach at any size: Start small, be consistent, and let results guide your next experiment.

METRICS AND RESOURCES

Use quantifiable measurements to justify investment and to choose channels on purpose. Social reach matters, but credibility lives on your own site.

Engagement metrics matter, but service metrics matter more.

Remember to track

- contact volume by issue before and after campaigns,
- first-contact resolution rates,
- repeat-call rates,
- portal adoption rates, and
- sentiment trends.

FIELD NOTES: DOS AND DON'TS

- Do communicate early, often, and honestly.
- Don't frame it as "educating" customers; better words are "engaging" and "communicating."
- Stick to one message across channels; tailor the tone, not the facts.
- Plan for accessibility: translations where needed, captions and alt text for all images.
- Make sure your communication—wording, images, or tone—reflects your entire service territory and community.

Consistency builds memory. If residents can repeat your "one thing" back to you, your message map is working. Plan for who might be left out by schedule, language, disability, or digital access, and close those gaps in your plan.

DROPS OF KNOWLEDGE

- Treat communications like infrastructure: plan, resource, and maintain it.
- Prevent confusion before it becomes contact volume. Tone influences compliance.
- Show your people and your work; faces and specifics build credibility.
- Go where customers are; publish natively and bring them back to your site for details.

- Measure a few outcomes that matter, not just reach; retire tactics that don't move the needle on the scale.

Great communication doesn't stop at service updates; it reaches the customer's bill through transparent conversations about cost, policy, and fairness. In the next chapter, we will dive into that communication topic.

A special thank you to Travis R. Thompson and Jimmy Luthye as well as the entire Denver Water team for participating in this chapter and sharing their story.

WORKS CITED AND BIBLIOGRAPHY

Adams, J. 2024. "Splashstreet Boys Spill Their Secrets to Success." *Denver Water TAP.* www.denverwater.org/tap/splashstreet-boys-spill-their-secrets-success (accessed March 2, 2026).

AWWA. 2022. *AWWA Utility Benchmarking 2022: Performance Management for Water and Wastewater.* Denver: AWWA.

Edelman. *2024 Edelman Trust Barometer Global Report.* Chicago: Edelman.

JD Power. 2023. *U.S. Water Utility Residential Customer Satisfaction Study.* Troy, MI: JD Power.

NBC News. 2024. "See Utility Company Denver Water's Funny Backstreet Boys Parody." *Today.* www.today.com/video/see-utility-company-denver-water-s-funny-backstreet-boys-parody-210043973957 (accessed February 27, 2026).

Pew Research Center. 2025. *Social Media and News Fact Sheet.* www.pewresearch.org/journalism/fact-sheet/social-media-and-news-fact-sheet/(accessed February 27, 2026).

Southern Nevada Water Authority. 2020. "Reality check—there are big fines if you don't change your watering clock." www.youtube.com/watch?v=wOll-ZBHM3dM (accessed March 2, 2026).

Town of Flower Mound Texas. 2025. "Give a Drip." www.youtube.com/watch?v=NkM5cXAWasQ (accessed March 2, 2026).

Chapter 17

From Cost of Service to Rates: Communicating the Value of Water

Water and wastewater utilities operate for cost recovery, not profit. Your community expects fairness, stability, and transparency. This chapter bridges two views: how costs are allocated (cost of service) and how those costs become customer-facing rates and policies, making it easy for leaders and frontline teams to explain rate changes with confidence.

MAKING VALUE FLOW

Municipal governments need to maintain certain fund balances to keep up their ratings with funding agencies. Most agencies hire outside help for this task; these rate experts have the tools to calculate rates needed to meet the expenses.

Rate Studies

Rate studies can form the foundation for a utility's annual operations. A rate study should have three principal objectives or components, described next.

Revenue sufficiency analysis. Conduct a revenue sufficiency analysis by developing and populating multiyear forecasting models for the city's water, wastewater, and solid waste systems to determine the level of annual revenue required to satisfy each system's projected annual operating, debt service, and capital cost requirements as well as to maintain a reserve fund balance at or above policy level.

Cost-of-service analysis (COSA). Complete a thorough cost allocation process to reflect the appropriate distribution of water system, wastewater system, and solid waste costs to customer classes based on observed usage profiles. The COSA allocates costs to system functions and then distributes functional costs to customer classes to allow for a comparison of current revenue collection by customer class with class-specific cost of service to enable the appropriate recovery of system costs from each customer class.

Rate design. Review the city's existing rate structures and develop modifications as appropriate to ensure that the city's rates conform to accepted industry best practices and reflect the current distribution of system costs while achieving policy objectives such as fiscal stability and proportional allocation of costs. The guiding tools for the rate studies are AWWA's M1, *Principles of Water Rates, Fees, and Charges*, and the Water Environment Federation's MOP No. 27, *Financing and Charges for Wastewater Systems.*

Each state and city has its own requirements. Examples of rate requirements could be to

- proportionally recover the utility's current cost of service and revenue requirements from each customer class;
- conform to accepted national and local industry best practices; and
- promote efficient water usage.

Per the last requirement, in US states where water scarcity is an issue, it is important to include water conservation tools in the rate structure.

While a utility sets its own schedule for conducting rate studies, a 2-year cycle is the standard recommendation. During years in which there is no rate study, the utility conducts a financial model update that reviews the effect of inflation and other trends. Rate adjustments are then recommended to address inflationary and other short-term budgetary factors, support long-term viability, provide resources for infrastructure maintenance and replacement, and align costs with the values and priorities of the city council and community.

Rate studies have become more accurate, thanks to technologies such as advanced metering infrastructure (AMI) data and better calculation tools that allow utilities to estimate future expenses. The AMI data set has become an important rate-setting tool by using the peak-use factors determined by AMI data for cost allocation. AMI expands the data set that can be analyzed from the previous 12 monthly data points per customer to 8,760 hourly data points per customer per year. This information incorporates more detailed demand characteristics of customers (average day, maximum day, and peak hour) and leads to results that reflect the actual observed usage characteristics of the utility's customers.

The AMI data set gives a utility the ability to understand customer-specific usage patterns and connect those patterns with the utility's costs to provide service. Peak-use patterns of customer classes contribute significantly to water system costs, thus providing an important consideration in developing class-based rates. Many municipal governments operate wastewater treatment facilities, so those expenses need to be evaluated in addition to water rates. Customer costs are generally allocated according to the number of accounts, and collection costs are commonly allocated on the basis of volume, but treatment costs can be recovered through charges.

Cost Estimations

Planning for the future is an important responsibility for a utility manager. And when it comes time to request municipal funding, especially additional funding to cover rising costs, it helps if the manager has established the city council's trust in the utility's services. This is done by providing a solid rate case that consists of comprehensive, accurate cost estimates, such as making sure to include fixed costs for service fees when estimating water and wastewater sales. Professional rate consultants are invaluable in helping a utility develop a long-term strategy. A rate study will assess the full cost for delivering services as well as the revenue required to maintain a utility's financial stability.

Rate Changes and Customer Service

Communicating with the public about rate changes is important in securing public support for a utility's decision (see Figure 17-1 for one example). Explaining and being transparent to customers on how the money is spent helps build trust.

Educate your utility's customer service staff by reviewing rate studies with them. This helps representatives discuss the studies confidently and knowledgeably with customers. Have both printed and online information available to meet different customer needs. Recognize any non English–speaking customers in your service area.

Rate changes in the billing system should be tested systematically to ensure that accurate rates have been entered into the billing system. To maintain customer confidence in your utility, be ready to explain rate changes and have tools for customers to express their concerns. Open meetings to present a rate study are a great way to do this. Invite customers to get involved and to provide feedback before the rates are adopted.

It's an essential aspect of good customer service to be able to admit mistakes. Something might go wrong during the rate-adjustment process. Notify customers of the error, and tell them how it has been resolved. With a solution in place, customers are more likely to accept a utility's mistakes.

Leak Adjustment and Customer Service

Most water utilities have a leak adjustment policy in place. When customers have received a large bill resulting from a leak, they're likely to turn to their utility for help. Customer service representatives (CSRs) will benefit from having the utility's leak adjustment policy on hand for these situations.

If a utility bills customers according to an inclining rate structure in which the rate increases with usage, the utility may allow a customer to pay at the lower-tier rate in the case of a leak. The customer is still responsible for payment, but the impact is lessened by the utility's solution. The utility's policy should specify for how many billing cycles and how often a leak adjustment is allowed. (AMI data and alerts can help customers navigate leak issues.)

Source: AWWA 2026

Figure 17-1 Cost of water relative to common products

CSRs are likely responsible for addressing customers' payment arrangements and payment plans. While guidelines and rules need to be in place, empowering representatives with the option to be flexible can help reduce the number of escalated calls and dissatisfied customers. The billing system should be able to keep track of adjustments, especially if the number of adjustments is limited. Representatives should keep clear, concise notes so that others working in the billing system are aware of previous incidents and arrangements.

DROPS OF KNOWLEDGE

- Explain the bridge: revenue sufficiency → cost of service → rate design.
- Use AMI and observed patterns to allocate costs fairly and prevent surprises.
- Convert analysis into clear policies and frontline talking points before going live with communications.
- Publish summaries in plain language and show corrections transparently.

Next: Customer service in the community—especially when a utility is making changes that directly affect its customers, a detailed, multifaceted outreach plan is essential.

BIBLIOGRAPHY

AWWA. 2026. *Customer Service Training Program. Course 3: Utility Industry Operations.* Denver: AWWA.

AWWA. 2017 (7th ed.). M1, *Principles of Water Rates, Fees, and Charges.* Denver: AWWA.

Water Environment Federation (WEF). 2018 (4th ed.). MOP 27, *Financing and Charges for Wastewater Systems.* Alexandria, VA: WEF.

Chapter 18

Community Intervention: Customer Service in the Street

Great customer service doesn't live only at a call center. It shows up where people are—at neighborhood meetings, after work, on weekends, and in the channels customers already use. Community intervention means showing up early, speaking plainly, and making it easy for customers to understand what's happening and take action.

Utilities ask communities to invest in aging systems and to adopt new tools that change how people interact with their water service. That means utilities are responsible for explaining what's changing, why it matters, and how customers can stay in control.

The American Society of Civil Engineers' 2025 Infrastructure Report Card continues to rate major water categories below average (including drinking water and wastewater), reinforcing the notion of why utilities must connect "what's changing" to reliability and reinvestment in terms customers can understand.

Not all customers are available during traditional hours, and some households still rely on limited broadband or smartphone-only access. In-person, after-hours touchpoints, and paper-friendly materials help close those gaps. The Federal Communications Commission's report on US internet connections underscores why "online-only" change management is risky. Even when service is available, subscription patterns vary by place and household, so utilities should still plan for paper-friendly materials, in-person help, and after-hours touchpoints, especially during billing and portal transitions.

Community intervention is a program, not a press release. It blends neighborhood meetings, multichannel reminders (including person-led calls and social posts), bill inserts and one-page factsheets, city council meetings, and hands-on support like onsite portal enrollment or bill review. The goal is consistent: Turn confusion into clarity and surprise bills into solvable stories.

METRICS AND RESOURCES

Ideas for Community Meetings

- Neighborhood community group meetings
- Public library sessions, like a once-monthly "Customer Service in the Street" meeting
- School district or local school meetings: e.g., using a "Meet the Teacher" format to reach residents whose child may be the primary English-translator in the home
- Financial assistance or low-income-taxpayer clinic events
- Church events and local fairs with "roving" customer service

Before launching a community program, equip your team with plain-language talking points, avoiding jargon that you and other utility employees might commonly use (e.g., *seamless*, *efficient*, *innovative*, *disrupting*). Be sure to accommodate all people, including those who may want to see illustrations or actual items; showing an advanced metering infrastructure (AMI)–enabled meter and transmitting device can be more meaningful than just talking about them. Civic meetings should involve visuals and talking. Take time at the beginning to explain any terminology that is specific to your presentation and have handouts available for people to peruse post event.

Time your community outreach so that residents hear from you before they see a change on their bill. Repeat the message across meetings, direct calls, website banners, and partner organizations.

A UTILITY TALE

Recently, a utility located in the northeast United States was planning a citywide switch from an older billing platform to an upgraded one. This involved a corresponding change to billing frequency (from quarterly to monthly) and a parallel AMI program. In addition, the aesthetic look of the bill and the water and sewer rate structure/pricing were changing, from a tiered water use model to a cost-to-serve model. Ahead of this overhaul, the utility led a full-scale community intervention campaign. These are the steps it took.

- Visits to city council member offices were held to explain the changes and allowed the council members to see what the new meters would look like and ask questions about the rates and how they would impact their residents.
- The utility held briefings with a simple script and slides at neighborhood community association meetings as well as statewide financial support conventions to make income-vulnerable customers aware of assistance programs and policies that would change during the implementation.

Hi, I'm [Name] with exciting news!

On X date the utility will go live with its new billing system, providing you with a more predictable, reliable, and customer-friendly water billing service.

Improvements include monthly billing and a Customer Service web portal with real-time access to your water usage. Community meetings, where representatives will explain the new billing system, are just getting underway. I encourage you to attend and learn more about what these improvements mean for you.

For a complete list of meetings and for more information, visit [URL] or call xxx-xxx-xxxx.

Thank you.

Figure 18-1 Robocall script example

- It made targeted robocalls during go-live week to reinforce monthly billing and direct residents to the website (see Figure 18-1 for an example).
- Customers received bill inserts and concise factsheets that explained fees, the new bill layout, and where to get help.
- All employees received small, pocket-sized cards with information that included phone numbers for customers to call if they had questions. This way, any utility employees who were approached, no matter what their role, could give a customer the card to provide consistent wording and information.
- The utility provided onsite portal sign-ups so customers could set alerts and start monitoring their use right away.
- TV spots included recognizable faces in the community to show how to log in online and understand visuals such as graphs.

The rollout campaign was successful. Here's why.

- Plain language and consistent messaging were used across every channel.
- The new bill and the online portal were actually shown to customers rather than just described.
- Evening and weekend events were held for working families and caregivers.

DROPS OF KNOWLEDGE

Don't ever underestimate the budget for, need for, and absence of communication. You will pay one way or another, either proactively with a smart approach and gathering your community together or reactively with

MEASURING WHAT MATTERS

- Reach and access: Attendance; percentage of events after 5:30 p.m.; weekend versus weekday mix; percentage of onsite portal enrollments.
- Understanding: Percentage of attendees who rate the bill as "easy to understand" post-event; top recurring questions logged.
- Adoption and prevention: New portal enrollments; leak alerts set; follow-up calls avoided (tag "resolved at event").
- Trust signals: Survey item—"The utility communicates proactively about changes" versus baseline.

an unsuccessful project (because no one understood what you did) and angry reactions from customers.

Create a reasonable timeline for your communications plan depending on the amount of change you expect from the project. The bigger the project, the longer the communications timeline should be.

Show, don't tell. Use plain-language scripts, visuals, and onsite portal sign-ups to turn confusion into control. Evidence from utility-bill redesign research shows that simplifying layout and presenting information in a more intuitive order can improve customer comprehension and satisfaction—exactly what utilities need when rolling out new bills, portals, or billing frequency changes.

Explain the money. Be transparent about fees and how they fund reliability; connect rate design to infrastructure needs. The US Government Accountability Office cites US Environmental Protection Agency estimates that more than $630 billion will be needed over 20 years to overhaul clean water and wastewater infrastructure, numbers that make it easier to explain why fees and project timelines matter.

Don't forget the employees. Making sure that even your internal stakeholders know the same track is important. People will often approach your staff in the streets or on a project site, so consistency is key.

Next: Apply the brand under pressure and when headlines hit, how to stay steady with the media.

BIBLIOGRAPHY

American Society of Civil Engineers (ASCE). 2025. "ASCE Report Card Gives U.S. Infrastructure Highest-Ever 'C' Grade: Stresses Need for Sustained Investment to Support Economic Growth." *Civil Engineering Source.* www.asce.org/publications-and-news/civil-engineering-source/society-news/article/2025/03/25/asce-report-card-gives-us-infrastructurehighest-ever-c-grade (accessed March 2, 2026).

Federal Communications Commission (FCC). 2025. *Internet Access Services: Status as of June 30, 2024.* https://docs.fcc.gov/public/attachments/DOC-411463A1.pdf (accessed March 2, 2026).

US Government Accountability Office (GAO). 2024. *Clean Water: Revolving Fund Grant Formula Could Better Target States' Varying Infrastructure Needs.* GAO-24-106251. Washington: GAO.

Chapter 19

When the Story Becomes the Story: Media, Perception, and Trust

You've probably seen this before: A TV news anchor and an elderly woman on a fixed income clutching a water bill. The reporter repeats the amount of the bill; neighbors and family members shake their heads. What news watchers don't hear: months of past-due balances layered on top of a small, undetected leak that grew; a meter that did its job; and a customer service department sending several notifications. The balance due is real, but the news story is incomplete because it lacks context and next steps.

Stories like this surface all the time. With social media megaphones—Nextdoor, neighborhood Facebook groups, TikTok clips—one segment can harden into a community belief about your utility before your team finishes its morning briefing. That first impression shapes how customers call, post, and show up at city council meetings.

This chapter is about meeting that moment. We will discuss how to prepare for media and social amplification and complete the story with empathy, plain-language facts, and credible partners, so that people know what happened, what to do now, and what your utility will do next.

HOW MEDIA SHAPES PERCEPTION (AND UTILITY WORKLOAD)

It's a matter of speed versus certainty. News and social platforms reward speed with a 24/7/365 news cycle, algorithms from social and analytic platforms reward clicks, and science rewards certainty. In water and sewer services, these sources and outcomes rarely align. Rumors, misunderstood data, or plain lies fill the gap, and repeated exposure changes the algorithms that can make a bold claim about billing or water safety feel true.

Consider visuals and framing. A photo of a glass of brown water or a bold headline can anchor belief more than a PDF of lab results. When the message is framed as "danger," audiences are primed to notice any confirming detail.

As John Medina, developmental molecular biologist, writes, "We are incredible at remembering pictures. Hear a piece of information and three days later, you will remember 10% of it. Add a picture and you will remember 65%" (Medina 2026).

Consider the echo effect. Influencers, neighborhood groups, and private testing kits can create herd mentality, where everyone believes that something is correct en masse. Even after a claim is disproven, adjacent worries (e.g., other contaminants) often surface. Plan for the second wave of questions.

Reality Check: Not Every "Neighbor" Online Is a Neighbor

In 2018, Facebook reported removing roughly 1.3 billion fake accounts across two quarters (Q4 2017 and Q1 2018). Company leaders publicly acknowledged the scale of removals that year.

In mid-2018, reports showed Twitter was suspending more than one million accounts per day, totaling more than 70 million accounts in May and June alone.

Why it matters to water: Some "local" posts that nudge people toward bottled water or amplify doubts about tap water may be seeded by bots or coordinated campaigns as opposed to real neighbors. Treat sudden surges and repeated talking points that were seen online as signals to lean on verified data, clear explanations, and partner voices.

A UTILITY TALE

Mt. Pleasant Waterworks, S.C.

The following case study shows how one utility navigated a monthlong media surge without losing its footing or its voice.

In midsummer 2017, several residents using home screening kits believed they had detected pesticides in their drinking water. Social posts spread quickly, and a national environmental advocate known for work in the energy space amplified the concern. Local and regional media followed the story closely. For more than a month, Mt. Pleasant Waterworks (MPW) faced intense public scrutiny and calls for action.

What MPW Did

- Engaged partners immediately: MPW coordinated with Charleston Water System and the state health department to align sampling and messaging.
- Sampled where people were worried: Crews collected samples at customers' homes and throughout the system, mirroring where the concerns originated.

- Used an independent, US Environmental Protection Agency–approved laboratory: Samples were analyzed by testing service Eurofins Scientific for a broad suite of analytes, including pesticides and herbicides (and the specific compound mentioned in community discussions).
- Publicly released results and held meetings: MPW shared findings with customers and media and fielded questions at community sessions.
- Stayed engaged when the story shifted: Even after results showed no detection of pesticides or herbicides, new questions arose about emerging contaminants, and the utility and partners addressed those directly.

What the Data Showed

Results reported publicly in late July 2017 indicated no pesticides or herbicides detected in the finished water. The analysis covered more than 200 compounds and included the pesticide specifically named in community conversations. As an added trust-building measure, MPW launched its Citizens Academy, helping to educate residents who were interested in the water cycle and who, as a result, became advocates of the utility.

Why It Worked

- Be first, be right, be credible. MPW responded early, centered verified data, and used third-party validation.
- Sample where the concern lives. Sampling at customers' taps acknowledged lived experience and reduced the "You didn't test my water" gap.
- Use one voice but multiple messengers. Alignment with neighboring systems and public health agencies reduced mixed signals.
- Offer transparency rather than spin. Sharing methods and lab results helped reset the message from rumor to evidence.

MPW's Lessons

- Treat social posts as early detection, not as adversarial attacks.
- Pair speed with humility. If you don't have an answer yet, tell when you will.
- Publish raw lab results with a plain-language explanation.
- Anticipate the second wave of questions (e.g., concerns about per- and polyfluoroalkyl substances, DuPont's GenX, color, taste, or odor).

FROM CONCEPT TO CULTURE: BUILDING MEDIA READINESS BEFORE YOU NEED IT

Defined roles. Name primary and backup spokespeople. Empower supervisors to escalate customer or media communication quickly. Codify who signs off on data and who posts updates.

Message maps. Pre-draft three-tier answers (10-second, 30-second, 2-minute) for your top 10 media scenarios: discoloration, low pressure, main break, boil-water advisory, taste and odor, emerging contaminants, rate changes, disconnections policy, cyber incident, and wastewater overflows.

Partner alignment. Keep memoranda of understanding in place with neighboring systems and health agencies. Share templates so that statements coordinate across logos.

Channels that fit your town. Maintain an updated landing page (e.g., "Water Quality Updates"), SMS/email list segmentation, and relationships with local reporters. In small communities, your Facebook page may reach more customers than a press wire.

Practice under pressure. Run a quarterly 60-minute drill for utility staff: drop a mock headline in the room and time your first public post, partner notification, and media callback.

THE PROOF IS IN THE BOTTLE: BOTTLED AND VENDED WATER

The Guardian investigated California's 9,200 water vending machines, noting many were placed in low-income, immigrant communities with markups of 8,000–10,000% over tap (Koran 2019). An Environmental Working Group report alleged the then-market leader exploited immigrants' fear of tap water.

The Voice of San Diego mapped water vending machine density; most were concentrated in tracts with lower incomes, more rentals, and higher shares of noncitizens and Latino residents (DiPerro 2024). Customers paid approximately \$0.30–\$0.50/gal for vended water versus approximately \$0.01–\$0.02/gal for tap water.

As early as 2011, Nestlé Waters North America's corporate communications director confirmed the company marketed bottled water to Black, Indigenous, and people of color communities (Whitman and IPS 2011).

Per Manuel Teodoro, a professor of public affairs, in his discussion with *Circle of Blue*,

> So, to drink tap water is to trust government [and] [t]o drink bottled water, or its commercial alternatives, is a sign that you distrust government because you're willing to pay tens to hundreds of times more for a commercial product that you believe is superior—even though bottled water is less stringently regulated. . . Who distrusts tap water the most? Based on the evidence, low-income households and communities of color. Those groups most frequently experience discolored, foul tasting, or contaminated water from their faucets. (Walton 2022)

Media Readiness Checklist

- Spokesperson roster with cell phone numbers
- 24/7 approvals path (two-deep)
- Prewritten advisories and holding statements
- Message maps for top 10 scenarios
- Dedicated updates page + SMS/email templates
- Media list (local, regional) + after-hours flags
- Communications memoranda of understanding partnered with contact tree
- Graphics kit (maps, treatment steps, "how we test")
- Social listening and rumor log
- Drill schedule + after-action form

COMMON PITFALLS (AND HOW TO AVOID THEM)

There are several items you should strive to avoid when working with the media. Don't be silent while you wait for perfect data. Post the process and the next update time while still avoiding jargon-heavy releases.

Avoid one-and-done posts. In a fast cycle, repetition is clarity. Plan a regular routine. Finally, make sure your partners also reflect your language, and if possible, use joint statements.

METRICS AND RESOURCES

- Time to first public update (target ≤60 minutes from awareness)
- Time to partner notification (target ≤30 minutes)
- Call deflection rate (spike in calls relative to outbound reach)
- Share of voice in local coverage (neutral/positive versus negative)
- Sentiment trend across owned channels (before, during, after the event)
- Contact with vulnerable customers (languages, accessibility, after hours)
- Customer understanding (post-event microsurvey on "knowing what to do")

DROPS OF KNOWLEDGE

- Speed + transparency earns trust. Be first with facts and next steps.
- Sample where concerns originate, and publish results plainly.
- Coordinate voices. Health agencies and neighboring systems are credibility multipliers.
- Plan for the next question. Rumors rarely end with the first lab report.

FINAL THOUGHT

You cannot control a headline, but you can control your readiness, your tone, and your follow-through. In a low-trust media climate, consistent, evidence-based updates delivered with empathy are your best defense and your best service.

Next: Back up the story with proof—how data from advanced metering infrastructure can explain bills, find leaks, and speed up fixes.

WORKS CITED AND BIBLIOGRAPHY

DiPerro, A. 2024. Voice of San Diego. "Everything You Need to Know About Water Vending Machines." https://voiceofsandiego.org/2024/12/13/everything-you-need-to-know-about-water-vending-machines/ (accessed March 2, 2026).

Koran, M. 2019. "Californians Are Turning to Vending Machines for Safer Water. Are They Being Swindled?" www.theguardian.com/us-news/2019/dec/02/california-water-vending-machines-quality (accessed March 2, 2026).

Medina, J. 2026. Brain Rule Rundown: Rule #10: Vision Trumps All Other Senses. https://brainrules.net/vision/(accessed March 2, 2026).

Rosen, G. 2018. "How Are We Doing at Enforcing Our Community Standards?" https://about.fb.com/news/2018/11/enforcing-our-community-standards-2/ (accessed March 2, 2026).

Timberg, C. and Tony Romm. 2018. "Twitter Suspends More Than 70 Million Accounts in Crackdown on Fake Users www.washingtonpost.com/technology/2018/07/06/twitter-is-sweeping-out-fake-accounts-like-never-before-putting-user-growth-risk/ (accessed March 2, 2026).

Walton, B. 2022. "Tap Water Failures and Distrust of Government: A Conversation with Manny Teodoro." *Circle of Blue*. www.circleofblue.org/2022/supply/water-management/infrastructure/tap-water-failures-and-distrust-of-government-a-conversation-with-manny-teodoro/ (accessed February 26, 2026).

Whitman, E. and IPS. 2011. "Bottled Water Companies Target Minorities." www.aljazeera.com/features/2011/11/22/bottled-water-companies-target-minorities (accessed March 2, 2026).

Chapter 20

Metering and Interval Data

Water utilities are finally embracing advanced metering infrastructure (AMI), also known as water smart-meter projects. For people who are not familiar, this project involves moving from traditional or drive-by read meters collecting data monthly to meters that can send and receive a signal and transmit reading data, usually hourly. Electric utilities did this 20 years ago and have moved to intervals more frequent than 15 minutes, but municipal governments have been financially pressured and, in some cases, reluctant or slow to adopt this technology that has made such a difference within customer service in water utilities.

AMI is not only a way to read meters, but it can also enhance and improve customer service. When a meter reader visits a customer location once a month to get the billable reading from a traditional meter, there is no way for the utility to know what happened in the past 30 days. When customers call the utility asking why their bill is so high, the water services professionals have no way of knowing what happened or when. AMI has changed this scenario. Research from the Alliance for Water Efficiency shows that AMI-enabled proactive leak notification programs significantly reduce the duration of continuous-flow leaks. In utilities that paired hourly interval data with customer outreach, leak duration dropped from multiple billing cycles to a matter of days, reducing both water loss and customer bill shock (Alliance for Water Efficiency 2023).

WHY IS AMI A GAME CHANGER?

Meter-reading automation is not new. Drive-by meter reading has been adopted by many utilities, yet they have been slow to move to the AMI technology.

Utilities strive to provide the best possible service, and AMI is a great addition to a utility's customer service toolbox. Imagine a customer calling to ask about their bill. The customer service representative (CSR) can provide detailed information: The customer's water usage increased starting on Friday the 12th at 10:00 a.m., this pattern lasted for 2 weeks, and the water usage was adjusted at 10:15 a.m. 2 weeks later. "*What happens on Fridays every two weeks?*" the CSR asks. Did the customer notice any changes? This conversation can help solve the mystery, especially if the meter-read data are shared with the customer. It will become evident that the customer's landscapers come every 2 weeks on Friday morning.

THE POWER OF DATA

Empowering Customers and Utility Staff

Data-driven conversations are powerful. Customer service improves, customer satisfaction increases, and utilities can manage their water waste. AMI is a win–win for all parties. Customer portals are great tools for sharing data with customers, and they can be customized to meet your utility's specific needs. Keeping customers informed in this way allows them to be in control of their water usage. They can set triggers, which are especially useful if a customer doesn't live within the utility's service area full time. Leak alerts can notify a customer of continuous use for a certain number of hours. Even if a customer isn't close to their home, they can be informed and take action. An evaluation of multiutility AMI leak alert programs found that customers who received timely notifications avoided thousands of gallons of water loss per incident. Earlier detection shortened mean repair duration and reduced the financial impact on households, especially in cases of irrigation or service line leaks (Alliance for Water Efficiency 2023).

CSRs are empowered when they have access to usage data; they can explain with numbers why a customer's bill is higher in a particular month. They can observe patterns, make year-to-year comparisons, and show how customer's usage patterns vary seasonally or how weather patterns affect water usage.

Plan for Seamless Customer Service

Meter data management is important after an AMI installation. Ideally, it is part of the customer information system (CIS); this saves CSRs the step of having to always sign on to multiple sites. Building connections between meter data management and CIS should be part of initial discussions. Be sure to involve customer service in developing and planning an AMI system.

Also, a utility should verify that the meters and endpoints are working and that installation records are correct. Even with a reliable system like AMI, it's a good idea to do occasional field checks, especially after a new installation.

With meter data management as part of the CIS, utility staff can see all meter data in the billing system without logging into another portal to view the reads: no need to create a work order to get a re-read from the field. Your utility creates one-stop call resolution by giving customers up-to-date information during a phone call and guiding them to see the data themselves. The

Water Research Foundation notes that utilities integrating AMI data directly into the CIS reduce repeat high-bill calls because representatives can resolve issues during a single interaction using interval data rather than dispatching field staff for manual re-reads (Water Research Foundation 2020).

Note that not all CIS systems are equally robust, so you'll want to evaluate the capabilities of your CIS, the degree of complexity of the interface from the AMI system, and the utility's long-term plan. If you want to stay with your systems and are happy with the CIS, it is worth investing in building the connections from the beginning.

An important partner in delivering AMI data to customers is your utility's water conservation division, if you have one. They will become your biggest advocates and help customers on a daily basis. They can train your CSRs to detect patterns and problems in customer consumption. They can train water customers how to use the portals and read the consumption data.

AMI IMPLEMENTATION: THINGS TO CONSIDER

- **CIS.** Is your billing system able to meet your needs today? Will it be able to adjust when you start collecting interval data? Can AMI data be inserted into the billing system?
- **Access.** Who needs access to the AMI data? How can you provide it to all parties, including customers? Who should not have access to AMI data? Policies around how external agencies can use data should be set up at the start of any project.
- **Upgrades.** When you upgrade any of the key systems, how can you prevent breaks between them? Is it best to connect the systems, or should they remain independent?
- **Training.** Don't limit AMI training to CSRs; provide this essential step to other utility staff, such as meter technicians in the field and water conservation–focused employees.
- **Use cases.** What use cases could be improved using interval meter data? Can you perform better financial forecasting? Water supply planning? While you may not know every use case, other stakeholders within your utility may.

If all utility employees understand how the AMI system works, they can create a successful story as they communicate with customers.

Communication

Any time you make changes to the systems that have a direct effect on your customers, communicate early and often. Change can be hard for customers and employees alike. Customers will want to know how the change is going, so build their trust during what could be an uncertain time for them. Keep customers engaged, remind them of the change that is coming, and focus on the positive impact of the change. Host workshops and maintain in-person communication. "Show and tell" can be invaluable; until a customer sees how AMI will help them, it is hard to visualize and understand the true benefits.

Employees may be stressed; they may worry about how this new AMI system will affect their jobs. Keep employees in the loop about progress being made with the AMI installation. Educate them. The role of the traditional water meter reader will be replaced by new responsibilities, such as managing the system by reviewing data or checking on no-reads and zero-use accounts. Assure these employees that they are essential to the long-term success of AMI. Defining the specific aspects of their job should be done early. New titles can be developed. Be sure to involve your human resources department from the very beginning.

AMI: MORE THAN A CUSTOMER SERVICE BENEFIT

So far, this chapter has focused on how AMI improves customer service. There are other significant benefits. AMI also strengthens nonrevenue water management. AWWA's guidance on water audits and loss control highlights how continuous-use flags and zero-use alerts improve leak detection speed and operational response time when compared to traditional monthly read cycles (AWWA 2024).

- **Rate studies.** AMI allows a utility to identify peak-use factors. What customer class has the greatest impact on the system? By identifying these customers through AMI data, a utility can assign the cost accordingly. AMI data show the peak-use day and hour, allowing a utility to custom-design its distribution system to meet water demands.
- **Water conservation.** Building a strong water conservation team usually comes after AMI is installed. You can start by deciding who will provide customers with what information and how much they will provide. CSRs and meter technicians will be the first responders. However, meter technicians are not conservation experts, so you may need a group of employees to focus on the utility's conservation efforts. These colleagues will know what options are available. They can do site visits and make recommendations based on those visits.
- **Jobs.** As mentioned earlier, the meter readers will see the biggest change in their jobs. Since they are no longer reading meters, clarify what are they doing on daily basis. Maybe retraining or new certifications are needed. Maybe CSRs' job descriptions will include new tasks. The CSR jobs will be more data driven and analytical than before. Update all job titles as required by the new AMI system. Work out all the details with your human resources department well before the job description goes live.

A UTILITY TALE

In 2016, Tempe, Ariz., briefed its mayor and city council on a plan to modernize water metering by moving from manual reads to an AMI system (City of Tempe 2016). At the time, more than 43,000 meters were read once a month by eight full-time meter readers to generate more than $80 million in annual water and sewer billing, an approach the city called labor-intensive and inefficient. The proposed AMI system would capture hourly reads from each meter and deliver a single billable reading to the utility billing system every month.

What Tempe built. The city outlined a fixed network with these core pieces: endpoints on each meter; repeaters and collectors mounted on structures like streetlights; and a meter data management system to store, analyze, and pass data to billing and staff/customer portals. Customers and customer services staff would be able to view historical consumption and request on-demand reads. Initial estimates called for about 50 collectors/repeaters citywide, designed with "2-to-1" redundancy so that every endpoint could reach at least two collectors. A vendor-led propagation study would finalize counts and locations.

Expected benefits of AMI. Tempe anticipated customer-facing gains (consistent billing cycles, improved accuracy, faster responses, early leak detection, and more transparency); operational gains (fewer manual reads and better system control); and financial and conservation benefits. Estimated annual offsets included $300,000 in labor savings, $30,000 in vehicle savings, and $100,000 from reduced losses, plus safety improvements and better preventive maintenance. Conservation features—like customer dashboards, early leak alerts, fewer truck rolls, and a smaller carbon footprint—were also emphasized.

How Tempe staged the rollout. After a council-approved pilot in fiscal year 2013–2014 tested three solutions over 6 months, Tempe secured capital funding and hired a consultant to define needs, shape the request for proposals, support evaluations, and facilitate implementation. A cross-department project steering committee provided oversight. Phase I installed the core hardware and software and 5,000 endpoints, providing data transmission to billing and on-demand reads. Phase II finished the remaining endpoints and network devices across the city. The AMI work was coordinated with the ongoing meter replacement program so that new meters were compatible with automated reading technologies.

Cost and bill impact. Tempe estimated the total program costs at $7.75 million over a 15-year life (hardware, software, and annual maintenance and support). With full deployment, the city projected the average customer's monthly bill impact at about $0.25, recovered within regular water and sewer charges.

Peer context. Tempe noted other utilities installing similar fixed-network AMI systems, including Scottsdale and Tucson (Ariz.), San Diego (Calif.), Baltimore (Md.), Cleveland (Ohio), Olympia (Wash.), and Madison (Wis.), among others.

Bottom line. Tempe's story is straightforward: Test the technology, phase the rollout, pair it with meter replacements, and translate operational savings into better service and a modest, clearly stated customer impact (City of Tempe 2016).

How can AMI change your utility, or how has AMI already changed it? Discuss and share how you see the benefits of AMI in your utility and in your job. New use cases can be discovered as you start these conversations with your colleagues in the utility and trade organizations. Sharing stories and information is essential in encouraging adoption of this innovative, truly revolutionary technology—no need to reinvent the wheel. Many have already gone through this process and can help you.

Next: Turn interval data into daily coaching, simple moves that cut demand without hurting service.

WORKS CITED

Alliance for Water Efficiency. 2023. *Smart Practices to Save Water: Evaluation of AMI-Enabled Proactive Leak Notification Programs.* Chicago: Alliance for Water Efficiency.

AWWA. 2024 (5th ed.). M36, *Water Audits and Loss Control Programs.* Denver: AWWA.

City of Tempe. 2016. Automated Metering Project Council Memo. www.tempe.gov/home/showpublisheddocument/43204/636076515904930000 (accessed March 2, 2026).

Water Research Foundation (WRF). 2020. AMI Meter Data Analytics: Guidance for Water Utilities. Denver: WRF.

Chapter 21

Water Conservation and Customer Service

Water conservation does not mean "using no water"; it means making intentional choices in policy, process, and daily behavior to use less. Scarcity pressures show up in drought cycles, in fast-growing communities that outpace supplies, and in aging systems that lose treated water before it ever reaches a tap. A utility and its customer service staff can help connect the dots for customers: why conservation matters, how your utility practices it, and what customers can do to help.

METRICS: WHY CONSERVATION NOW?

- Since 1999, average indoor household use has declined about 22% per household (≈15% per capita) as a result of efficient fixtures, proof that behaviors and technology move the needle (Water Research Foundation 2016).
- Utilities increasingly integrate nonpotable reuse for irrigation and industry as part of supply portfolios (US Environmental Protection Agency 2026).
- Conservation pricing and outreach must be designed with revenue stability in mind. Pair with transparent communication (California State Water Resources Control Board 2018, Pacific Institute 2012).

WHAT YOUR UTILITY CAN DO (OPERATIONS FIRST)

Start with your utility's own "house": secure and protect sources; operate production and treatment efficiently; maintain storage and distribution to minimize losses; and where relevant, reclaim and reuse water safely. Asset management isn't just about infrastructure; it's important to make sure a utility is operating optimally and that preventive maintenance is being budgeted and performed. Finance and billing play a role in water conservation, too, since rate design and customer programs can encourage smart use without surprising customers.

Source of Supply

Plan for and invest in reliability, from water rights and reservoir planning to wellhead and watershed protection. Forecast demand with conservation, reuse, and climate variability in mind.

Water Production and Treatment

Operate wells, pumps, and treatment trains to minimize waste. Where filtration is used, recycle filter backwash to the head of the plant when it's permitted and cost-effective. Track energy per million gallons and recovery rates to spot efficiency gains.

Water Storage and Distribution

Maintain tanks, towers, valves, and pressure zones to deliver water reliably while guarding against losses. Prioritize leak detection and timely repairs; verify meter accuracy and set a regular replacement cycle. Capital plans should reflect continuous improvement of the network rather than fixing things only when they're broken.

METRICS: SYSTEM EFFICIENCY

- Distribution performance: real losses (gallons/mile/day), apparent losses (% metering error), and data validity scores (AWWA water audits).
- Peak day-to-average day ratio: high ratios often indicate outdoor demand opportunities.
- Portal adoption: percentage of active accounts enrolled; leak alert resolution: median days from alert to fix.

Collection Systems and Water Reuse

In many communities, reclaimed water keeps parks green, cools power plants, and supports industry, freeing up potable supplies for drinking and hygiene. Understand how your utility or even the next neighborhood upstream discharges and reuses flows and how those flows influence the next utility downstream.

METRICS: WATER REUSE

- Common nonpotable uses: landscape irrigation, industrial cooling, construction, and habitat support (Environmental Protection Agency).
- Track reuse deliveries (in million gallons per day) and offset of potable demand; report customer outages avoided during peak-demand months.

BUSINESS OPERATIONS (RATES, BILLING, CUSTOMER PROGRAMS)

Increasing-block (conservation) rates can align price signals with efficient use when tailored to local conditions and paired with education. During drought, temporary surcharges may be used carefully to manage peak outdoor demand. Whatever you do, explain the block rate concept clearly to customers and show them how to stay in the lowest block for essential indoor needs.

As time passes, the idea of time-of-use pricing for water isn't so substantial; if your costs increase during peak periods as a result of your energy pricing, then shouldn't pricing be reflective of that change?

METRICS: PRICING AND DEMAND

- Price matters: Residential demand responds to price, with stronger effects where scarcity is acute.
- Design carefully: Conservation pricing should balance efficiency and revenue stability; adjust and communicate annually (Pacific Institute 2018).

CONNECT OPERATIONS WITH CUSTOMER SERVICE

Customers experience one utility. Build a strong connection between field, treatment, billing, and the call center so that conservation facts and fixes are consistent. Customer service should know the "why" behind notices, leak alerts, irrigation schedules, and rate structures and be able to explain them plainly.

EDUCATING AND INVOLVING CUSTOMERS

Use multiple channels, such as bill stuffers and SMS/email for alerts, a web hub with plain-language guides, and in-person workshops. Pair indoor rebates (toilets, washers) with outdoor programs (xeriscape conversions, irrigation checks). If you have advanced metering infrastructure (AMI) in place, help customers enroll in your AMI-enabled portal so they can see their hourly use, set alerts, and self-diagnose leaks.

METRICS: AMI AND PORTALS

- Water Research Foundation and utility case studies recommend linking AMI data to portals with customer-set alert thresholds and comparative usage views (Water Research Foundation 2020).
- The Alliance for Water Efficiency's recent evaluation found limited published savings data to date for leak-alert programs, underscoring the need to measure outcomes locally (Alliance for Water Efficiency 2023).
- Track % portal enrollment, # leak alerts closed, gallons saved from verified fixes, and repeat-leak rate.

A UTILITY TALE

Seattle Public Utilities (SPU) and 18 neighboring water providers formed the Saving Water Partnership. This collaboration includes tools, guidance, and rebates to help residents and businesses use water wisely.

The goal was to reduce per-capita demand despite population changes. The Saving Water Partnership combines education and outreach such as free gardening classes, technical assistance, and financial incentives that help customers fix leaks, upgrade fixtures (e.g., high-efficiency toilets), and improve irrigation. Through rebates and advisory services, the goal was for customers to use less water and still receive great service.

Seattle Public Utilities encourages customers to use meter readings for early leak detection, comparing "1 cubic foot" dial positions before and after a period with no water use to identify hidden leaks and reduce prolonged loss (SPU 2026a).

SPU also offers billing adjustments for most leaks if repairs are completed and inspected within 60 days, supporting innovation in customer-utility collaboration to limit extended water loss (SPU 2026b).

Seattle experienced approximately 30% savings below its highest point, and it has been one of the lowest consumption rates per person in the nation.

Bottom line: Using a coordinated approach with multiple utilities, setting a measurable 10-year goal yet still an achievable one, and putting their money behind it, the Saving Water Partnership made saving water a habit.

DROPS OF KNOWLEDGE

- Conservation starts with your utility's operations; protect sources, optimize treatment, maintain storage, and fix leaks fast.
- Use conservation pricing and plain-language education to help educate customers about water demand.
- Make AMI data useful; portals, alerts, and simple leak-find guides reduce waste and customer complaints.
- Track outcomes that customers feel: gallons saved, leaks fixed, avoided high bills.
- Partner with schools and community groups to build long-term conservation habits.

Programs work best when tools keep improving.

Next: from an intelligent voice assistant to predictive analytics and how to use them without the hype.

WORKS CITED

Alliance for Water Efficiency. 2023. *Smart Practices to Save Water: An Evaluation of AMI-Enabled Proactive Leak Notification Programs.* Chicago: Alliance for Water Efficiency.

California State Water Resources Control Board. 2018. State Water Board Drought Year Water Actions: Conservation Water Pricing. www.waterboards.ca.gov/drought/pricing/ (accessed March 2, 2026).

Pacific Institute. 2012. Water Rates: Conservation and Revenue Stability. https://pacinst.org/wp-content/uploads/2013/01/water-rates-conservation_and_revenue_stability.pdf (accessed March 2, 2026).

Seattle Public Utilities (SPU). 2026a. "Detect Leaks." www.seattle.gov/utilities/your-services/water/metering/detect-leaks (accessed March 2, 2026).

SPU. 2026b. "Leak Adjustments." www.seattle.gov/utilities/your-services/accounts-and-payments/bills-and-payments/my-bill-seems-too-high/leak-adjustments (accessed March 2, 2026).

US Environmental Protection Agency (USEPA). 2026. National Water Reuse Action Plan. www.epa.gov/climate-change-water-sector/national-water-reuse-action-plan-and-program (accessed March 2, 2026).

Water Research Foundation (WRF). 2016. Residential End Uses of Water, Version 2. Denver: WRF.

WRF. 2020. AMI Meter Data Analytics. Denver: WRF.

Chapter 22

Predicting the Future: GenAI, Bots, and Everything on the Way

"No Mail, Low Morale."

That was the motto of the 6888th Central Postal Battalion, the only all-Black, all-female unit deployed overseas during World War II. Their mission was straightforward yet essential: keep the mail moving. Without communication, morale among the troops collapsed.

The same principle holds true in today's water and wastewater industry: Communication is the lifeline between a utility and its customers. With physical mail having been overtaken by email and digital channels, mail now might come to customers in the form of a chatbot response, a text message, or a predictive notification about a high bill. Nevertheless, reliable, clear, and timely communication is still essential in maintaining the morale of utility staff and customers. Technology may advance, whether through artificial intelligence (AI), chatbots, predictive analytics, or voice assistants, but unless leadership, organizational change management, and customer response are considered right alongside those tools, utilities could be faced with low morale both from employees and their service base.

Adding to the challenge, the pace of change is faster than ever (see Table 22-1 for details about how quickly technology is changing). Even looking just 5 years ahead, it's difficult to predict which technologies will become essential and which will fade away. That uncertainty should shape how utilities approach projects. Large-scale initiatives that take 3–5 years to implement are increasingly risky because by the time they launch, the technology may already be outdated. Instead, utilities will benefit from adopting a mindset of continuous process improvement, investing in smaller, incremental steps that can evolve quickly. This approach allows utilities to adapt as technology advances while keeping customers at the center of the process. This shift also depends on the support of regulators such as city councils, county boards, commissions, and federal organizations who influence how utilities can finance these investments.

Table 22-1 Industry snapshot: artificial intelligence (AI) adoption across water and wastewater utilities

Year	AI Adoption Across Utility Functions
2022	~10–15% exploring or piloting AI in operations or customer service
2023	~15–18% adoption—billing tools, predictive leak detection, early chatbots
2024	~20–22% estimated expansion into customer self-service and analytics
2025	~25–30% using AI in at least one core function
2026 (Projected)	Potential 40–60% adoption across multiple areas
2035 (Projected)	Estimated 70–80% adoption across multiple operations

Sources: Xylem 2025, Murphy and Ours 2024

As of this printing, the treatment of investments that move to an approach of software as a service by local legislators has not been aligned with the desire to keep utilities secure and adaptable to the changes in service delivery and customer expectation. Moving toward a model of shorter, adaptive technology projects requires regulatory flexibility and a shared understanding that continuous improvement ultimately strengthens service reliability, customer trust, and financial stewardship.

THE NEW TECHNOLOGY LANDSCAPE

Generative AI

Generative artificial intelligence (genAI) refers to systems that can generate responses, create text, or provide explanations that are human-like. In customer service, this means answering questions about bills, explaining policies, or even walking customers through conservation programs. While genAI cannot replace the human touch, it can help reduce call volume and provide faster responses for simple requests. By drafting, authoring, and using relevant context, genAI can reduce the number of impersonal or confusing form letters as well as the time it takes for clean water professionals to manually draft correspondence for many customers about the same topic.

Here's a simple example. A customer types into a utility's online chat, "Why is my bill higher this month?" Instead of waiting for a representative, a genAI tool might instantly check usage history and reply with, "Your water use was 25% higher than last month, and based on the amount and timing of your water use, it is likely an indoor fixture like a toilet. In addition, you had a period of 3 days during which your water was continuously running. Warmer weather and outdoor irrigation may explain the increase, but check fixtures in your house as continuous use of water could have created this situation."

Intelligent Virtual Assistants

Intelligent virtual assistants (IVAs) use speech recognition, summarization, or what's called sentiment analysis to help customers through phone systems. Instead of navigating a menu by pressing numbers, customers can simply say, *"I want to pay my bill"* or *"Report a water leak."* For many utilities, this technology shortens call times and frees staff to handle more complex situations. It also helps customers avoid, by using natural language, the ever-frustrating decision over what keywords can be used to have a virtual assistant understand them.

Bots and Automated Responses

Bots are automated tools such as chatbots on a website, text bots for billing reminders, or even social media bots that provide quick updates. Their job is to handle routine questions. However, bots are only as effective as the information behind them. If poorly designed, they can frustrate customers and increase rather than decrease call volumes. They can include genAI services or be as simple as standard FAQs.

Predictive Analytics

Predictive analytics uses data to anticipate problems before they happen. For operations, this could mean spotting a water main likely to fail using sensor data. For customer service, it might mean predicting which accounts are at risk for late payment so that staff can contact a customer early. Most importantly, predictive analytics helps customer service and asset maintenance groups improve their response to customers in terms of availability of water service or potential interruptions.

Sentiment Analysis

Sentiment analysis looks at the tone, or emotional content, of customer communications like phone calls, emails, or social media. When used carefully, sentiment analysis can highlight at-risk customers and give utilities a chance to intervene early with empathy and personalized solutions. But note the potential gray area in which helpful segmentation can become unfair profiling. The goal is not to label customers but to listen more effectively and respond with respect. Ideally, using sentiment analysis for agent training and potential process issues is a great use case for this technology.

WHAT NEW TECHNOLOGY MEANS FOR CUSTOMERS

These emerging technologies share a common goal: to make the customer experience more effective (through understanding a bill or learning about local utility projects), more personalized, and more reliable. They also shift expectations. Once customers get used to getting instant answers from their bank or internet provider, they expect the same from their water utility.

A UTILITY TALE

Las Vegas Valley Water District (LVVWD) built its customer-facing experience around an IVA to make self-service faster and to free up agents for complex needs. The IVA guides callers through common tasks and proactively alerts customers to account changes. It's a digital front door that connects people to the right help without long menus.

What makes this effort stand out is the closed-loop learning behind the scenes. LVVWD analyzes call transactions and voice recordings for sentiment and performance signals. Those insights drive rapid improvements to the IVA, tweaking call flows and adding new intents so customers reach the right path more quickly. Instead of waiting months for a system release, small changes roll out continuously to keep the experience aligned with real customer behavior.

Training the IVA is treated like training a team. The utility began with a broad library of intents, each seeded with diverse phrases (e.g., at least 20 phrases per intent across hundreds of intents), so the system would recognize the many ways customers ask for the same thing. Now, missed-intent reports and sentiment results inform ongoing updates, improving accuracy and reducing dead ends.

LVVWD also monitors "mood insights" to pinpoint friction and coach people and processes, not just the bot. Examples include refining queue routing versus intent training, troubleshooting login issues, and addressing payment-application frustrations. Actions range from enriching intent phrases and adjusting routing to targeted agent coaching and, when necessary, escalating fixes to outside vendors. When the signal shows a positive experience, the team avoids unnecessary changes. In short, the same telemetry that fine-tunes the IVA is used to train agents and improve upstream processes.

This is continuous improvement in practice. The cycle—listen, learn, and adjust—reduces contacts that either have not taken place or have moved to the least expensive channel, supports agent training, and keeps customers moving. The benefits stack up over time: cost savings, better quality, higher employee engagement, stronger customer satisfaction, and improved productivity. Just as important, the approach should stay flexible. Small, frequent improvements prevent the platform from going stale rather than waiting for big multiyear upgrade projects.

THE RISK OF PREDICTING TOO FAR AHEAD

The truth is that this book cannot predict the future, and we will not be so bold as to say with certainty what customer service will look like in 5 years. A few years ago, hardly anyone was talking about genAI. Now it's one of the most discussed topics in every industry. Similarly, technologies that seem essential today may fade tomorrow. The lesson is to prepare for change.

According to a recent report by scholars at MIT NANDA, many AI projects have not yet delivered meaningful revenue or cost improvements at scale (Challapally et al. 2025), so we can derive from this that people and their timeless skills are just as important as the technology they implement.

WHEN TECHNOLOGY GOES TOO FAR

Technology is powerful, but it is not a cure-all. Sole reliance on drafted answers without validation (also known as *workslop*), ever looping yet agreeable bots, and not using all the necessary variables to make a customer's experience with the technology meaningful leads to

- frustration when bots or automated systems fail to understand unique customer situations,
- exclusion when less tech-savvy customers are left behind, and
- data misuse and unethical AI when predictive or sentiment models cross the line from helpful insight to unfair profiling.

The lesson is balance. Technology should enhance, not replace, human connection. The most successful utilities will use technology wisely without losing sight of the human relationship at the core of customer service.

DROPS OF KNOWLEDGE

- Technology is changing fast. Utilities must adapt but not lose sight of customer trust.
- Sentiment and predictive analytics must be used carefully. There's a fine line between personalization and profiling.
- Balance matters. Technology can speed up service, but human empathy builds long-term confidence.
- Keeping your system current is about flexibility. Utilities that stay nimble will be best prepared.
- Think in smaller, faster steps. Long, multiyear projects risk becoming outdated. A continuous improvement approach, supported by regulators, boards, councils, and leadership, keeps utilities responsive to customers and change.

FINAL THOUGHT

The 6888th Central Postal Battalion proved that communication sustains morale: "No Mail, Low Morale" was more than a motto; it was a truth. For today's water and wastewater utilities, the lesson is the same. Technology can be a powerful enabler, but without leadership, organizational change management, and continuous improvement, it cannot carry the weight of customer trust on its own.

Utilities that move in smaller, adaptable steps supported by regulators and guided by a clear focus on customers will be best positioned to keep morale strong, both within their organizations and across the communities they serve.

Next: technology helps, people deliver—how roles evolve and why humans remain essential.

WORKS CITED

Challapally, A., C. Pease, R. Raskar, and P. Chari; MIT NANDA. 2025. *The GenAI Divide–State of AI in Business 2025.* www.aigl.blog/state-of-ai-in-business-2025/ (accessed February 27, 2026).

Murphy M., and J. Ours. 2024. Why AI Adoption for Utilities Lags Behind and How to Catch Up. https://centricconsulting.com/blog/why-ai-adoption-for-utilities-lags-behind-and-how-to-catch-up_ai/ (accessed February 27, 2026).

Xylem. 2025. "25% of the World's Public-Sector WWTPs to Use AI in 2025." www.xylem.com/en-us/support/lets-solve-water-blog/25-of-the-worlds-public-sector-wwtps-to-use-ai-in-2025/ (accessed March 2, 2026).

Chapter 23

Why New Technology Will Not Eliminate the Need for People

Technology is now part of how every utility delivers service from lab testing, plant inspections done by drones, digital assistants, dashboards informed by advanced metering infrastructure, leak alerts, online portals, and automated drafting tools.

With each new technology solution, it's easy to wonder: If chatbots can answer questions, digital voice technology sounds human, and systems write letters, what's left for people to do in customer service?

Plenty.

New tools change the tasks, but you and the team of clean water professionals you work with still drive the outcomes for the customer: trust, understanding, and fair resolutions. Automation removes much of the manual work that you may have already secretly hated and gathers data faster (no more copy/paste errors), so naturally there is a fear that comes with the idea that there may be fewer people needed in tasks. What technology has continued to show us time and time again is that there is no replacement for empathy and cross-department coordination. Even the best artificial intelligence (AI) needs humans to design, monitor, improve the decisions being proposed or approach to take, and step in when situations are novel, sensitive, or high stakes.

Take the example of medical specialists reading an MRI (magnetic resonance imaging). One doctor can only read the number of MRIs they are exposed to over a lifetime. Large language and image models can take in an infinite number of MRIs and spot patterns. In an odd parallel, while the act of reading an MRI may be initially performed by a large language model, there are no documented cases of a hospice nurse being replaced in a patient's home in favor of a bot during end-of-life care. Clean water professionals are providing the empathy that is needed to drive trust in very technology being used to power safe, reliable service.

Take a breath. As policies evolve and new rules around the use and misuse of AI continue to evolve, it isn't a battle between people and technology. In customer service, the balance tilts toward timeless human skills even as tools become more sophisticated.

Timeless Skills

How do these timeless skills show up in service?

- Providing context and unspoken iterations and flow because these skills are not one-size-fits-all. Neurodiversity and other factors may not be assessed effectively by technology when using sentiment analysis or when the situation is emotionally charged or confusing.
- Coaching customers through behavior change: e.g., how to read usage graphs, what a leak alert means, and where to start. Using technology to answer questions that are based in fear of technology (like smart meters) is not the best approach. Using the very people who use the data to help explain it grows trust.
- Owning exceptions. When policy, technology, and equity considerations collide, the decision is a human one.

Technology in the Customer Service Center

From a classic customer service representative's view, the toolkit keeps expanding: phones, customer information systems, bots, AI, digital assistants, email, website links and QR codes, and online bill pay, with new tools always on the horizon.

Phones still matter. Customers who feel heard, are called by name, and hear their name pronounced correctly come away from the conversation with more than an answer; they leave with dignity. Automation can triage and speed up routine calls, but it can't replicate a genuine human connection.

Writing is changing, too. AI can draft inserts, FAQs, notices, and web content. But AI doesn't eliminate roles; it shifts them. People curate tone and accuracy, adapt messages for different audiences and languages, and ensure alignment with policy, legal, and equity commitments. In many centers, senior agents are becoming practical AI modelers, training tools on the nuances they have learned from thousands of conversations.

Technology in Field Services (Metering and Maintenance)

AMI and analytics reduce manual reads and unnecessary truck trips, but they do not eliminate field work; they redistribute it. The work moves toward diagnostics and communications: splicing and testing endpoints, triaging network issues, and using interval data to pinpoint problems quickly. Two-way communication and early leak detection expand what field teams can accomplish. The result isn't fewer people; it's people doing higher-value work, like outbound communication with customers or making sure that service levels around reconnections can be done with better information and with the salary to map to those new requirements.

Technology in the Hands of Customers

Online bill pay and customer portals are now table stakes. Customers now expect these features to be available but did not 20 years ago. Many utilities also offer usage-monitoring apps that notify customers about unusual patterns such as sprinkler leaks, running toilets, or other anomalies. These have largely become commonplace, but they are not as widely used as once anticipated.

These programs still rely on people to design and promote the offering, enroll customers in conservation or rebate programs, produce training materials, and interpret alerts when customers call. Self-service reduces friction but isn't always accessible to all; guided service using the channel the customer chooses builds understanding and trust.

A realistic workforce outlook: Some scripted entry roles will shrink, yet hiring remains steady, given the number of retirements and promotions. Rather than the disappearance of customer work, you can expect more specialized roles, data analysts replacing the traditional "call handler" role, and the presence of digital care coaches, as well as experience and affordability managers.

METRICS AND RESOURCES

- AI is mainstream and additive. Organizations reporting generative AI use reached approximately 71% in 2024, spreading across multiple business functions (McKinsey & Company 2024). This use accelerates the need for people who design and govern usage.
- Jobs are changing, not vanishing. The US Bureau of Labor Statistics projects only a 5% decline in employment for customer service representatives from 2024 to 2034 yet estimates approximately 341,700 job openings per year due to replacement needs (US Bureau of Labor Statistics 2024).
- Call centers still need humans. Automation takes routine work, while humans handle high-stakes, identity-sensitive, and complex cases; even highly automated organizations have rehired for these functions.
- Water utilities see digital opportunity from within the lens of people. AWWA's *State of the Water Industry* and Black & Veatch's water reports show strong interest in using data and digital tools to improve decisions and service, efforts that require trained staff (AWWA 2024; Black & Veatch 2024).

DROPS OF KNOWLEDGE

- People + tech wins: Let automation handle repetitive tasks, so staff can focus on teaching, triage, and trust-building.
- Redefine roles rather than removing them: Provide training ladders into AMI, analytics, and digital care.
- Design for exceptions: Make the automated path fast, and make human handoffs graceful and timely.
- Govern the content: Treat AI drafts as first drafts; people own tone, accuracy, and local alignment.
- Invest in data literacy across teams: Reading interval data points and explaining them to customers are now core skills.
- Mind the hiring pinch: Use digital assistants to reduce strain on staff while you retrain for new roles.

We end here on purpose. Technology won't replace you or your people, but it will reward utilities that can reshape roles, equip teams with new skills, and re-center customer understanding.

The future of great service in the clean water industry remains humans amplified by tools, with empathy and building public trust leading the way.

WORKS CITED

AWWA. 2024. *State of the Water Industry Report.* www.awwa.org/state-of-the-water-industry/ (accessed February 26, 2026).

Black & Veatch. 2024. *2024 Water Report.* www.bv.com/en-US/resources/2024-water-report (accessed February 27, 2026).

McKinsey & Company. 2024. *The State of AI in Early 2024.* www.mckinsey.com/capabilities/quantumblack/our-insights/the-state-of-ai-2024 (accessed February 27, 2026).

US Bureau of Labor Statistics. 2024. *Occupational Outlook Handbook: Customer Service Representatives.* www.bls.gov/ooh/office-and-administrative-support/customer-service-representatives.htm (accessed February 27, 2026).

www.ingramcontent.com/pod-product-compliance
Ingram Content Group UK Ltd.
Pitfield, Milton Keynes, MK11 3LW, UK
UKHW020656041026
15590UKWH00001B/297

9 781647 172916